REBECCA EWING

REBECCA EWING

Basic Drafting For Interior Designers

Basic Drafting For Interior Designers

William E. Miller

Associate Member American Society of Interior Designers,
Board of Directors Designers Lighting Forum

VNR VAN NOSTRAND REINHOLD COMPANY
NEW YORK CINCINNATI TORONTO LONDON MELBOURNE

Library of Congress Catalog Card Number: 81-21880
ISBN: 0-442-26178-0
ISBN: 0-442-26177-2 pbk.

Manufactured in the United States of America

Published by Van Nostrand Reinhold Company Inc.
135 West 50th Street, New York, N.Y. 10020

Van Nostrand Reinhold Publishing
1410 Birchmount Road
Scarborough, Ontario M1P 2E7, Canada

Van Nostrand Reinhold Australia Pty. Ltd.
17 Queen Street
Mitcham, Victoria 3132, Australia

Van Nostrand Reinhold Company Limited
Molly Millars Lane
Wokingham, Berkshire, England

15 14 13 12 11 10 9 8 7 6 5 4 3 2 1

Library of Congress Cataloging in Publication Data
Miller, William E. (William Ernest) 1950–
  Basic drafting for interior designers.

  Includes index.
  1. Drawing—TAchnique.   2. Interior decoration—
Designs and plans.   3. Interior decoration in art.
I. Title.
NC825.I45M5   747′.022′1   81-21880
ISBN 0-442-26178-0   AACR2
ISBN 0-442-26177-2 pbk.

October 5, 1981

William Miller's book, *Basic Drafting for Interior Design* has captured the enthusiasm of our students and faculty. Though each group appraises the contents from a different point of view, jointly they concur that the approach is clear, concise and thoroughly absorbing.

Mr. Miller's own classroom experiences have shaped his direction. He is extremely aware of the need for an enriched text that both embodies the essentials of a formal background, as well as includes the concepts of imaginative creativity. His knowledge of interior design and drafting is extensive, and his ability to "Put it all Together" will help interior design departments pass on to their students an excellent teaching tool.

Sincerely,

Terry Reynolds
Assistant Dean
Parsons School of Design
Program Director of Parsons Midtown Campus

FOREWORD

All too often, creative people and students entering the interior design field have difficulty communicating their concepts. To help develop and advance their ability to plan space, to design, and to develop presentation skills, it is necessary to know how to draft. William Miller's "Basic Drafting Handbook" is structured in a simple, easy to understand manner. He is literally taking the participant "by the hand", in a clear step-by-step process through all the drafting elements involved in the production of an effective visual presentation.

For those seeking membership in a professional organization, a qualifying examination is obligatory. "The Basic Drafting Handbook" will prove invaluable to those professionals who wish to "brush up" on their forgotten drafting skills, as well as the young designer preparing for the examinations.

As a professional interior designer as well as an instructor of interior design, I am so pleased, at long last, to have this book as an aid to studio instruction. I know the students will benfit greatly from its application.

Thank you Bill for feeling the need.

Charlotte Saper
Member of the Board of Directors
New York Metropolitan Chapter
American Society of Interior Designers

PREFACE

Drafting is a skill that can be learned by almost everyone and may be developed into a talent by some. It is not part of our genetic or psychological make-up, programmed by nature, but rather it is an additional resource available to us, permitting the release and freedom of expression of our inner design thoughts. Interior design is a trade that is not only aesthetically oriented, but is technically oriented as well.

Although most of the principles and concepts laid forth in the text of this book are equally applicable to the field of drafting for architects, it does not, nor attempts to, cover the realm of material that is unique to an architect's jurisdiction. Structural load capabilities, plumbing, coding laws, and the like are definitely the domain of the architect, and an interior designor would be well advised to consult an architect if the project on which he is working requires such expertise. The draftsperson's job is to vivify a most scientifically exacting set of drawings from a compilation of sketches, calculations, specifications, descriptions, and intangible concepts.

Drafting is the vehicle by which our creative genius and design fantasies are born, the vehicle by which we gain entry into the world of reality. Conversely, it permits the transformation of that which we perceive in a three-dimensional and tangible world into one of two dimensions and non-existence. It is a skill that is called upon when some form of exacting communication, leaving no chance for misunderstanding or error, is required between two or more people. Through drafting, we formulate our ideas and designs, and we convey our dreams to those around us for execution, in order to ultimately gain appreciation and admiration for the transformance of our concepts into reality.

—WILLIAM E. MILLER

INTRODUCTION

The purpose of this book is of an introductory nature. Its contents are basic and its presentation simple. It is a book written specifically for those who have no drafting experience, for those students who wish an additional or simplified study aid, and for those professionals who now find it necessary to expand the skills required in an ever increasingly demanding trade, namely, interior design. There is no particular drawing style advocated in this book, for each person has within him or herself his or her own style, and must develop it to a comfortable working and professional level. What is provided here are guidelines that are considered somewhat universal in the world of graphic presentation. This is to insure a crisp, clean, understandable, and interesting drawing. That, in addition to the introduction of the graphic tools one employs in order to accomplish this, is the intention of this book. These are the qualities necessary to command the proper attention each project is due.

All illustrations of plans have been kept to a very basic level in order to permit fullest comprehension of the supporting text. It is therefore important to bear in mind that more intricate plans do exist and are quite commonly used in the field. To repeat, this is a basic but comprehensive introductory drafting handbook for interior designers.

As mentioned, the presumption is made that the student has no knowledge of drafting. Therefore, the sequence of chapters is arranged so that one can build upon the knowledge acquired in preceding chapters. A relationship is therefore established that will facilitate further understanding of ideas, suggestions, knowledge, and formats set forth in subsequent chapters.

When applicable, a review, noting the most important points covered in each chapter, has been included at the end of each chapter. These points should be kept in mind constantly and reread to permit one to use the information given in this book to its best advantage.

In addition, to further reinforce those ideas set forth in this book, a list of quiz questions has been provided. These questions relate to the information given in the previous chapter. They will prove to be useful as a guide to determine how much information the reader has retained.

Tables of standard size furniture, windows, doors, and kitchen appliances have been provided in order to acquaint the reader with universal measurement of standard items, and serve as a reference guide.

Exercises have been suggested at the end of appropriate chapters so that the student has an opportunity to apply in a practical situation the theory that he or she has learned. In doing so, future theory will have a stronger foundation for its application.

A glossary has been incorporated into this book so that the student may have immediate access to any definitions of unfamiliar terms that are either drafting or architecturally oriented.

In addition, choices for additional reading have been suggested, along with this author's impressions of each selection mentioned. The list of additional reading is provided for the reader who wishes to expose him or herself to a more in-depth coverage of each area of drafting mentioned in this text. Also, reading material has been suggested that covers topics related to drafting for interior designers, such as one- and two-point perspective drawing, construction techniques, rendering, paraline, axonometric, and orthographic drawing, and the science of ergonometrics.

By using this suggested reading material, one may become more aware of the relationship of drafting and the surrounding scientific arts associated with interior design.

Careful consideration and forethought have been given to both the choice of each selection and the author's comments so that the reader is made aware of those additional aids that will be of most use at the proper time.

CONTENTS

Basic Drafting For Interior Designers

1
TOOLS OF THE TRADE AND THEIR PURPOSES

Use of the proper tool is of the utmost importance whether speaking in terms of drafting, sports, or the culinary arts. For instance, one would not use a catcher's mitt in order to hit a home run, nor try to boil water in a broiler. It therefore follows that each drafting instrument has a particular function and should be used specifically for that purpose for which it was designed. Tools are aids at our disposal that can be effective only when utilized properly. Keeping this in mind, tools will work for and not against us.

DRAFTING SURFACES

For a proper drafting surface, a drafting board or drafting table is used. Sizes may vary, standard sizes being 12″ x 18″, 18″ x 24″, 24″ x 36″, and 30″ x 42″. (See illustration #1.) Forty-two inches is the recommended length. However, the size of the board to be used may be dictated by the size of the project on which one is working.

T-SQUARE

A T-square is used to draft horizontal parallel lines across a drafting surface whenever a parallel rule is not used. A T-square may also be used as a guiding support for other tools. (See illustration #2.) T-squares are made out of different materials—metal, wood, or plastic. Metal is usually considered the best material for T-squares since it does not warp, as will wood or plastic. However, plastic-edged T-squares permit the draftsperson to see through to the paper on which he or she is drafting. This capability is of paramount importance as it facilitates one's job immeasurably for obvious reasons. But the selection of T-square type is a decision which must be made by personal preference.

In addition, T-squares come in various sizes—2′–6″ long, 3′–0″ long, 4′–0″ long, etc. T-squares also come either calibrated or non-calibrated. Calibrated T-squares are similar to rulers and have inch measurements and breakdowns indicated. Calibrated T-squares, however, are considerably more expensive than non-calibrated, and quite frankly do not warrant the expense for our purpose. In any case, the T-square should be the same length as your drafting board.

drafting Table

drafting board

ILLUSTRATION 1

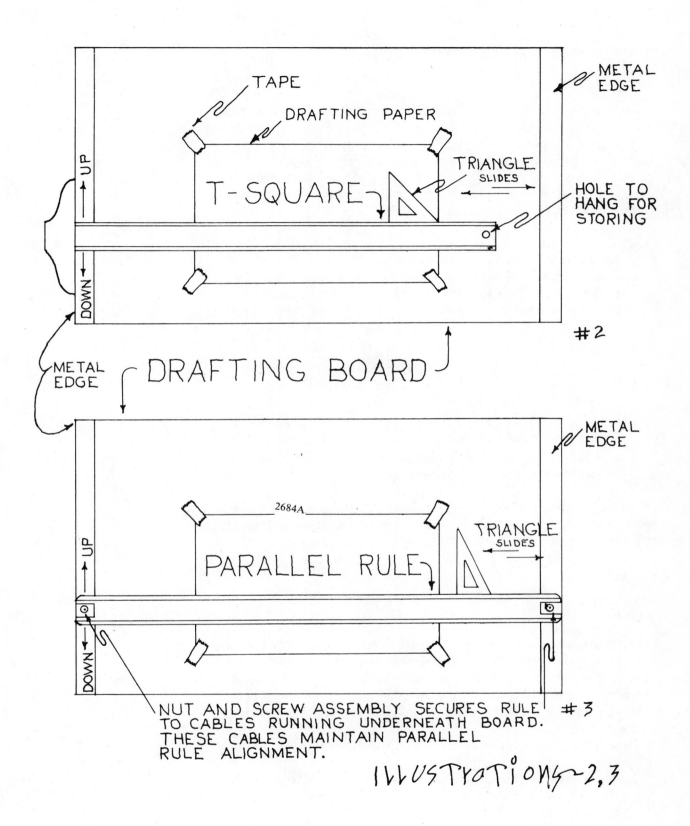

TAPE

DRAFTING PAPER

METAL EDGE

UP

T-SQUARE

TRIANGLE SLIDES

HOLE TO HANG FOR STORING

DOWN

#2

METAL EDGE

DRAFTING BOARD

METAL EDGE

2684A

UP

PARALLEL RULE

TRIANGLE SLIDES

DOWN

NUT AND SCREW ASSEMBLY SECURES RULE TO CABLES RUNNING UNDERNEATH BOARD. THESE CABLES MAINTAIN PARALLEL RULE ALIGNMENT.

#3

ILLUSTRATIONS 2,3

3

PARALLEL RULE

Some drafting boards come equipped with a parallel rule, which enables you to draw parallel horizontal lines quickly and easily. It is also used as a guide for other instruments. This will be explained later in this chapter. If a drafting board or table does not come equipped with a parallel rule, a T-square can be used to obtain parallel horizontal lines (See illustration #3.); however, there are certain drawbacks to using a T-square that make a parallel rule preferable:

1. A T-square is not attached to the drafting board or table, as is a parallel rule. Therefore, a greater chance of error in drawing true parallel lines is evident, since one end of the T-square may move or slip. This can cause problems and errors on the drawing that may not be noticed until much later, after having spent a great deal of time and effort.

2. Only one edge of the T-square can be used for drawing; both edges may not necessarily be exactly parallel.

3. Because one end may waver or fluctuate, only the other end may be used to cut paper (see section on paper).

4. When it is necessary to transport your drafting material, a T-square becomes an additional tool that needs to be carried.

5. A T-square cannot slide up and down your drafting board as smoothly and as fast as a parallel rule.

Although a drafting board or table with a parallel rule is slightly more expensive than one without, it is well worth the expense. I would advocate using one, especially for the beginning draftsperson.

PAPER

As mentioned in the foreword, each tool should be used only for the specific purpose for which it was designed. Consequently, each of the several types of tracing papers should be used only for the purposes intended for each. Slick papers are intended for ink, whereas a paper with more "tooth" (quality of paper to hold lead) is best used when lead is the medium.

Paper grades and their respective uses are as follows:

1. *Sketch or Rough Draft Grade.* Lightweight, least expensive, usually white or yellow, purchased by roll or pad, tears easily.
Example: 24″ x 50 yd roll, Charrette, white trace, 905. Under $10.00.
Use: Rough drafts, quick sketches.

2. *Medium Grade.* Fine to medium tooth, reasonably priced, white color, purchased by roll or pad.

Example: 24″ x 20 yd roll, Charrette, white trace, 120R. Under $10.00 ·

Use: Preliminary sketches, general lay-out work.

3. *Quality Grade (Vellum Paper).* 100% rag (crisp to the touch), expensive, purchased by roll or pad.

Example: 24″ x 20 yd roll, Charrette, 916H. Under $20.00. 18″ x 24″ pad with 50 sheets, Charrette, 916H. Under $20.00.

Use: Finished drawings, presentation of quality work.

4. *Films.* Plastic-like in appearance and touch, expensive, purchased by roll or sheet. Single matte may be inked on one side, double matte may be inked on both sides.

Example: Any manufacturer. Sheet price for 24″ x 36″ sheet is under $3.50 for single matte and $7.00 for double matte. A matte is a usable side.

Use: Overlays, clear reproduction, permanent drawings. To be used with ink only.

Any of these papers are available in an art supply store, under varying manufacture names.

Be sure paper is in alignment with drafting surface so that all related edges are parallel. This can be achieved by using the edge of your drafting surface and the edge of your parallel rule or T-square as guides. (See illustration #4.)

In order to cut paper to size when using rolls of paper as opposed to a pad, a smooth straight edge can be achieved by measuring the length of paper desired, placing it under a T-square or parallel rule and holding it down while tearing the edge of the paper against the T-square or parallel rule. This must be done in one very fast movement or the result will be an uneven tear. (See illustration #5.) Paper, however, does come in what is considered somewhat standard or universal sizes. These sizes are indicated by a letter and have the following relative dimensions:

SIZE	DIMENSIONS
A	8½″ x 11″ or 9″ x 12″
B	11″ x 17″ or 12″ x 18″
C	17″ x 22″ or 18″ x 24″
D	22″ x 34″ or 24″ x 36″
E	34″ x 44″ or 36″ x 48″

When purchasing paper, one should request the size by dimensions and not by letter size. In doing so, one can avoid unnecessary confusion as to discrepancy of size.

DRAFTING TAPE

In order to hold drafting paper down securely on the drawing surface, drafting tape is used. Drafting tape, very much like masking tape in

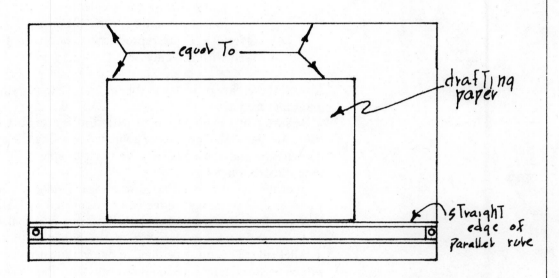

The edge of The T-square or parallel rule can be used To line up paper on drafting board.

illusTraTion ~ 4

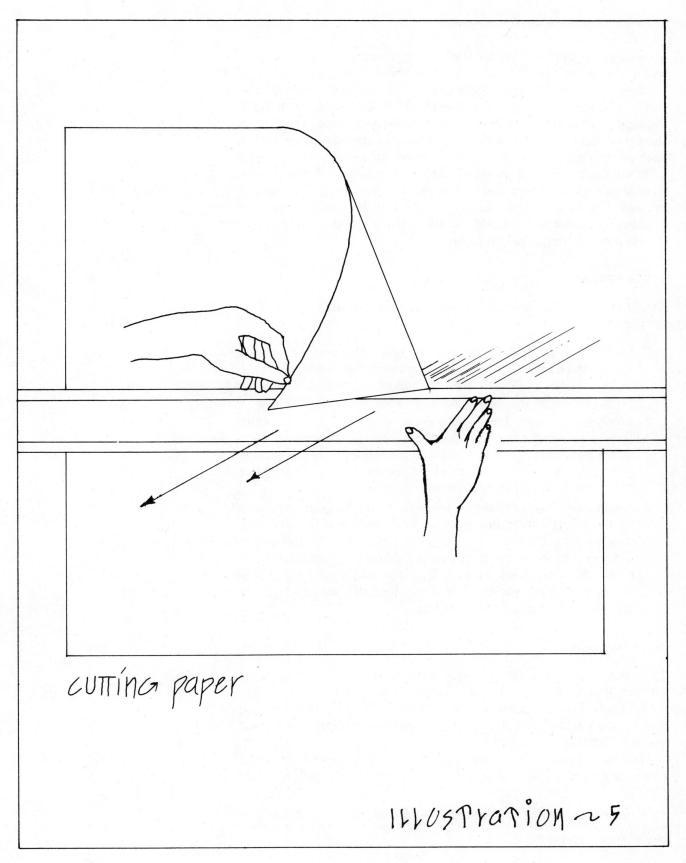

cutting paper

ILLUSTRATION ~ 5

appearance, is less sticky. It has the holding power of masking tape, but will not tear the drafting paper when you want to remove the tape. Drafting tape comes in several size widths and lengths (for example, ¼″ width, ½″ width, ¾″ width, etc.). One should choose the width of drafting tape most appropriate for the size of paper being used. The smaller the paper, the thinner the size of tape. Drafting tape should be used in each of the four corners of a drawing. (See illustrations #6 and #7.) This will create an even pressure from all directions and will prevent wrinkles or waves, which can distort the drafting paper and make it impossible to draw a straight or even-weight line. Be sure that the sheet of paper is parallel with the edges of the drafting board or table so that the finished drawing is not on an angle.

TYPES OF PENCILS

The following are common types of lead drafting or writing instruments.

1. *Ordinary Lead Pencil.* This may be used, provided the wood is shaven back about ¾″ to allow for proper sharpening; however, this tool is less preferable than the following types as it does not permit the flexibility of varying leads. (See illustration #8.)

2. *Mechanical Pencil.* Use a mechanical pencil with a .5 mm lead. It does not require frequent sharpening, provided one masters the technique of rotating the tool while it is being used. It is especially good for very fine lines. For thicker lines, however, a series of fine lines must be drafted. (See illustration #9.)

3. *Leadholder.* This is the most commonly-used type of pencil due to its flexibility. Either fine lines or thick ones can be produced, depending on the user's level of proficiency. It is not difficult to use, but it requires practice in proper sharpening and rotating. A leadholder uses a thicker type of lead than does a mechanical pencil. It is strongly preferred that the beginner use this type of instrument.* (See illustration #10.)

Pencil Technique

Two factors determine the technique one uses when holding a mechanical pencil or leadholder: comfort and practicality. The obvious slight variations in each person's technique may be attributed to individuality; that is, many techniques are acceptable, provided that they work well for that particular individual.

There are, however, some established guidelines that facilitate the proper use of a mechanical pencil or leadholder. These will alleviate

---

*Both the mechanical pencil and leadholder have a button on the top of the instrument which when depressed opens the bottom of the pencil barrel and permits insertion or exposure of lead. (See illustrations #11 and #12.)

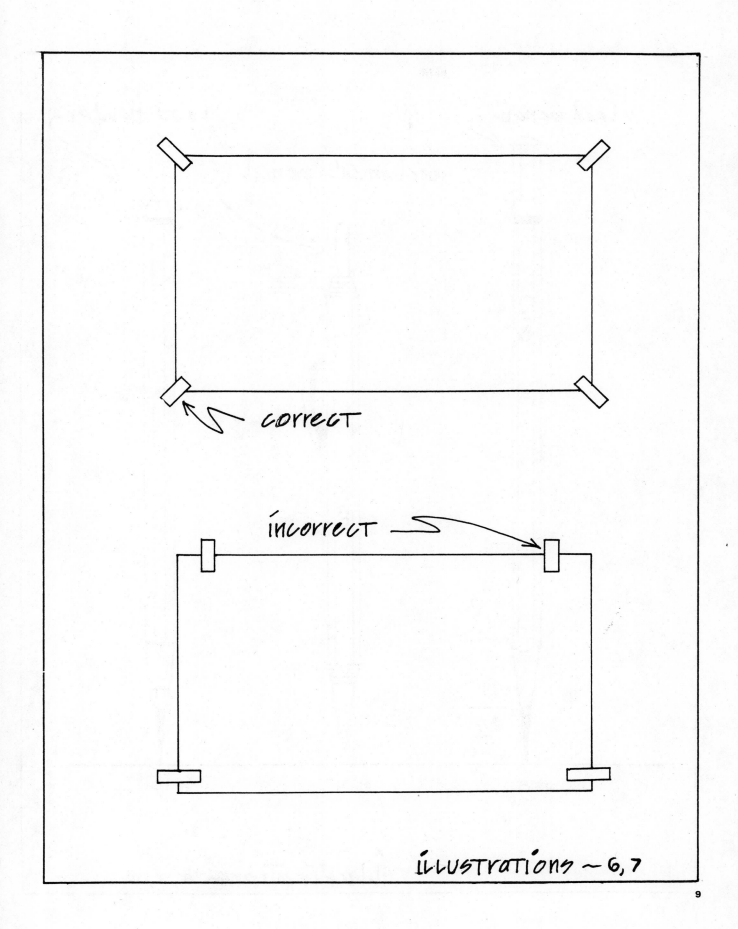

correct

incorrect

illustrations ~ 6, 7

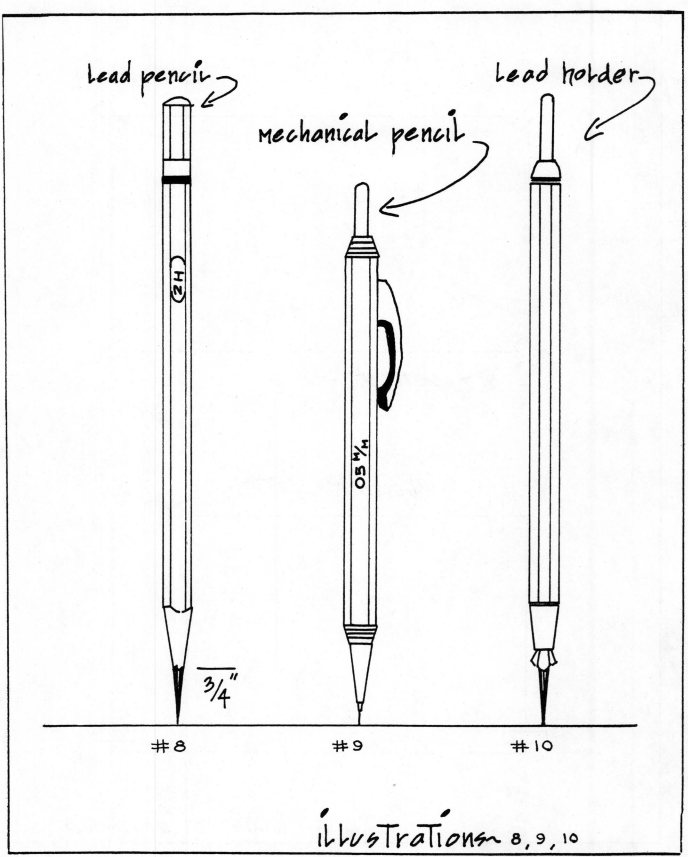

lead pencil

mechanical pencil

lead holder

3/4"

#8   #9   #10

illustrations 8, 9, 10

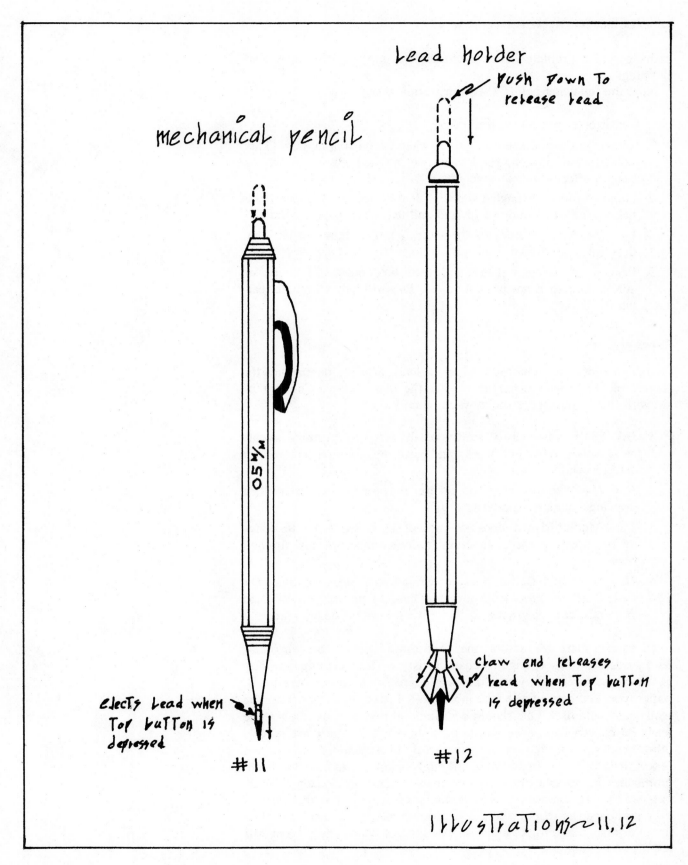

mechanical pencil

lead holder

push Down To release lead

0.5 M/M

ejects lead when Top button is depressed

claw end releases lead when Top button is depressed

#11

#12

ILLUSTRATIONS 11, 12

11

many common problems, such as smudging, paper tearing, and lead skipping.

Note the following points. (See illustration #13.)

1. Consistently sharpened lead point;
2. Rotation of pencil between fingers when pencil is in motion. (This technique will prolong the longevity of a pencil point and help to assure uniformity of line weights.)
3. Hand comfortably angled and slightly elevated above paper. (This helps eliminate smudging of lead and tearing of paper.); and
4. For right-handers, lines are drawn from left to right. For left-handers, lines are drawn from right to left.
5. Pencils, mechanical pencils, or lead holders should be pulled across the paper, not pushed across. This will help eliminate tearing the paper.

Types of Lead

Although there are numerous grades of lead (a grade determines the degree of hardness: the harder a lead, the more faint a line will be drawn), the following are the most commonly used:

1. *HB*—Soft: erases easily, smears easily. Used for primary layouts or sketches, dark line work, and lettering, requires control for detail work, blueprints well.
2. *F & H*—Medium: used for general purposes, good for finished drawings, layouts, and lettering.
3. *2H*—Medium Hard: does not erase easily if user has a tendency to be heavy-handed, hardest grade permissible for finished drawings.
4. *4H*—Hard and Dense: does not erase easily, does not blueprint well. Used for precise layouts, too hard to be used for finished drawings, creates grooves in paper if used with a heavy touch.

Of course, there are factors other than those inherent to a particular lead grade that may change its characteristics. These include humidity (an increase in humidity will increase degree of hardness of a lead), paper type, and paper finish (the more tooth a paper has, the harder the lead one should use). One should bear in mind that although only some leads have been mentioned (those used most often), there are several other grades available—ranging from 6B (exceptionally soft) to 9H (exceptionally hard). In addition, although specific leads have been recommended for specific purposes, one should experiment, using a different lead than that specified, as individual pressures of each draftsperson may influence the reaction of a lead. Remember, know what is called for in a particular project and use that type of lead! Any project will require more than one lead type.

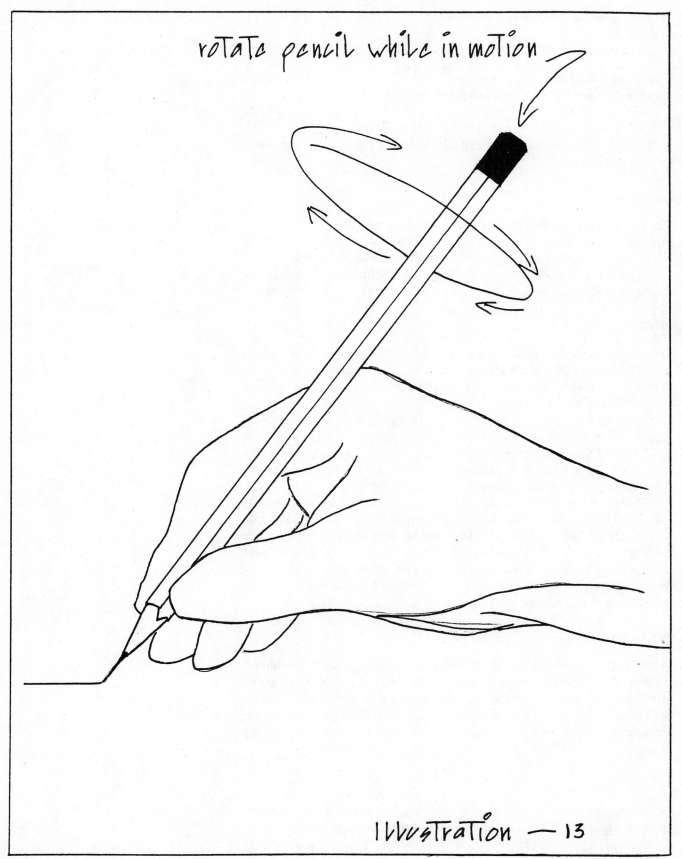

ILLUSTRATION — 13

Types of Points

There are basically two types of points one may wish to achieve when sharpening one's lead. (See illustrations #14 and #15.)

    A —Normal needle point (used for normal drafting)
    B —Chisel point (used for embellishment of lettering)—explained in later chapter.

Sharpeners

There are several ways to sharpen leads. The easiest way is to use a mechanical sharpener. (See illustration #16.) A sandblock, however, works just as well after one has had ample practice in sharpening the lead properly. (See illustration #17; see illustrations #18 to #21 for methods of sharpening lead for either a needle point or chisel point.)

PENS

Drafting or technical pens require a great deal of precision for proper use. They are not for the beginner, and one should master the use of a pencil or leadholder before moving on to a pen. Therefore, little time is devoted in this text to pens.

Technical pens come in various forms, depending on each manufacturer's specifications. Just as there are several types or grades of lead, one will find there are several interchangeable pen points. These points vary from #5 x 10 (which produces an exceptionally fine line) to #6 (exceptionally broad line). (See illustration #22.)

A beginner's pen set should include at least four different points—3 x 10, 2 x 10, 1, and 5. Because ink has a tendency to dry quickly, one should always clean each point thoroughly after each use. There are cleaners on the market made specifically for this purpose. Soft or mild detergent may also be used when diluted with water. It is strongly recommended that pens be stored in an upright position (point facing upward) so that clogging does not occur.

Pen Technique

When using a drafting pen for drawing, many of the same techniques employed for pencils apply. A major difference, however, is that one must keep the pen moving swiftly across the paper so that ink blots do not occur. If the pen is permitted to remain on the drafting surface on a particular point for a prolonged period of time, an excessive amount of ink escapes from the pen, thereby causing the ink to blot.

Ink

Pelican Fount India is an excellent non-clogging ink. For other appropriate inks, check with your local artist's store. Not all inks are compatible with technical pens.

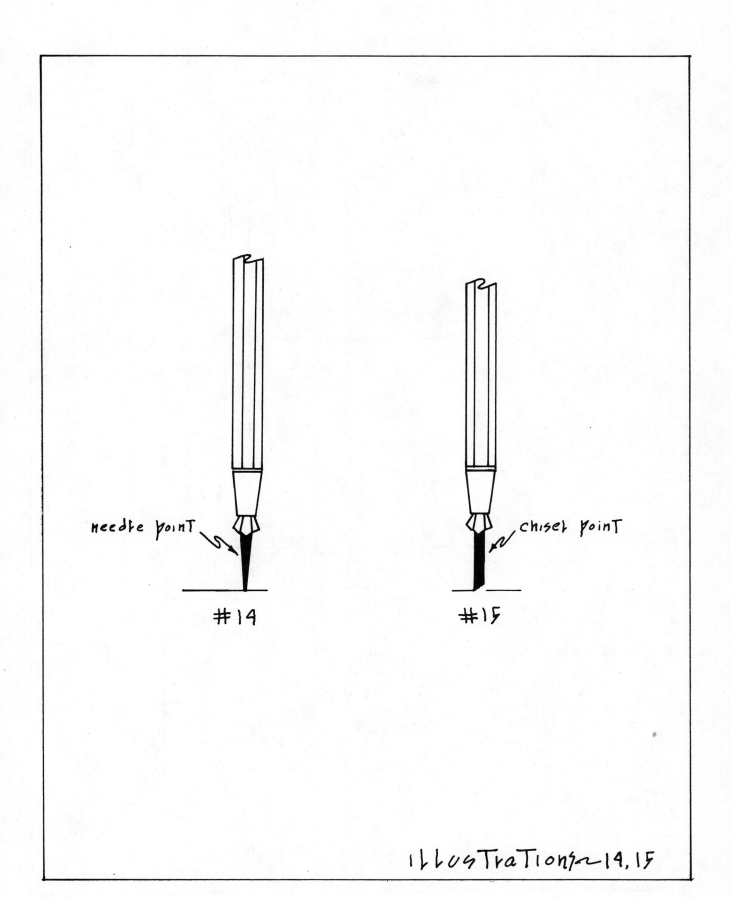

needle point

#14

chisel point

#15

ILLUSTRATIONS 14, 15

15

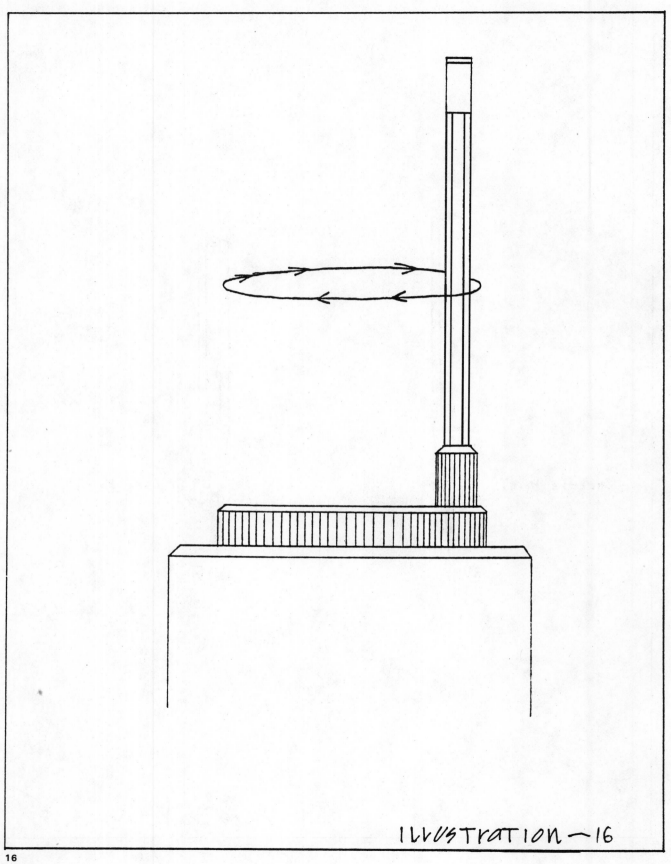

ILLUSTration—16

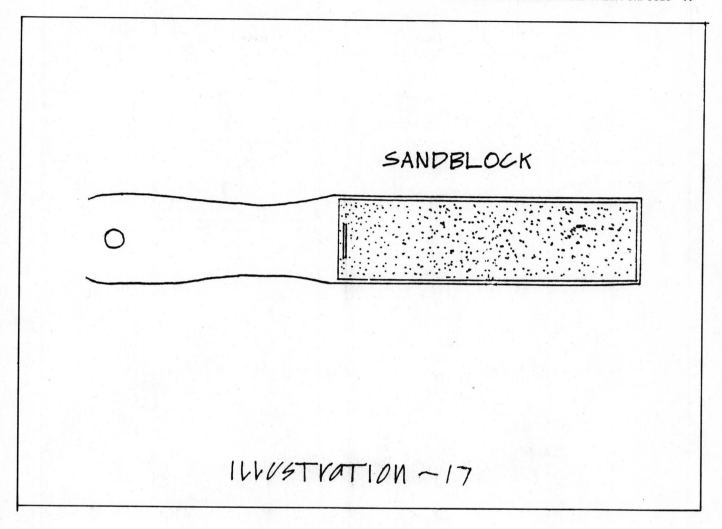

SANDBLOCK

ILLUSTRATION ~17

ERASERS

When using erasers, the goal is not only to erase, but to exercise care in order not to mar the drawing surface in the process. Erasers can tear, smudge, and otherwise destroy a good drawing. In order to avoid marring the surface, it is always advisable to use the softest eraser possible. Recommended erasers include Pink Pearl, Mars Plastic, Mars Staedtler, Magic Rub, and Faber Castell. Ink erasers are generally too abrasive and should be avoided if possible. If one is using ink or a mylar film, there are special liquids formulated to cover mistakes. In addition, there are electric erasers that one may use; however, these erasers do require careful attention when erasing, as they are extremely fast and will tear the paper. (See illustrations #23 to #33.)

Skum X is an erasing powder in a bean-bag that will erase smudges and smears over large areas of drawing. (See illustration #34.) Skum X also comes marketed under other names or may be in a container form. The container form is shaken over the drawing and then gently

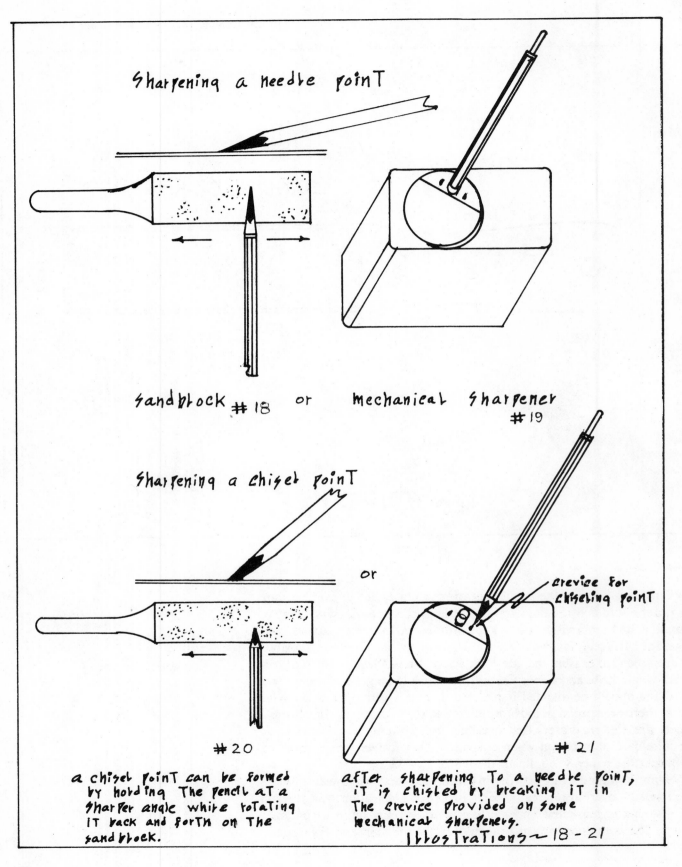

Sharpening a needle point

sandblock #18   or   mechanical sharpener #19

Sharpening a chisel point

or

crevice for chiseling point

#20

a chisel point can be formed by holding the pencil at a sharper angle while rotating it back and forth on the sandblock.

#21

after sharpening to a needle point, it is chisled by breaking it in the crevice provided on some mechanical sharpeners.

illustrations~18-21

Technical pen

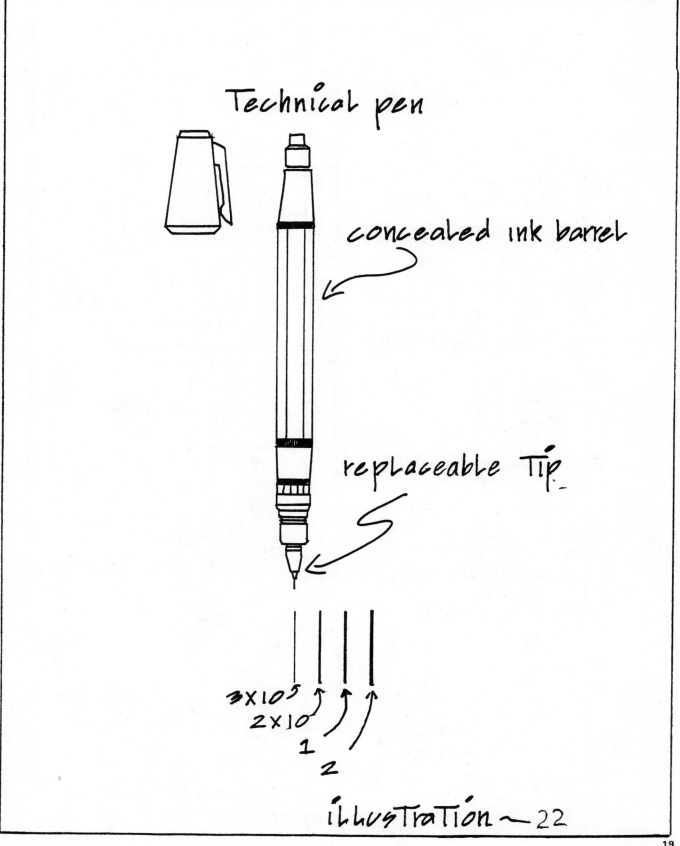

concealed ink barrel

replaceable Tip

3×10^5
2×10
1
2

illusTraTion ~ 22

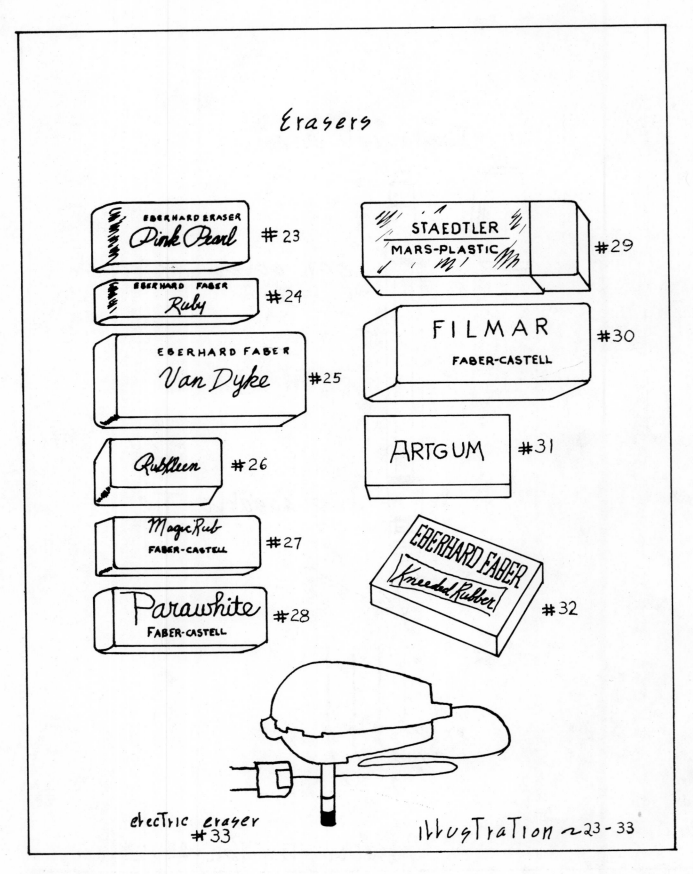

Erasers

EBERHARD ERASER
Pink Pearl #23

EBERHARD FABER
Ruby #24

EBERHARD FABER
Van Dyke #25

RubKleen #26

Magic Rub
FABER-CASTELL #27

Parawhite
FABER-CASTELL #28

STAEDTLER
MARS-PLASTIC #29

FILMAR
FABER-CASTELL #30

ARTGUM #31

EBERHARD FABER
Kneaded Rubber #32

electric eraser
#33

illustration 23-33

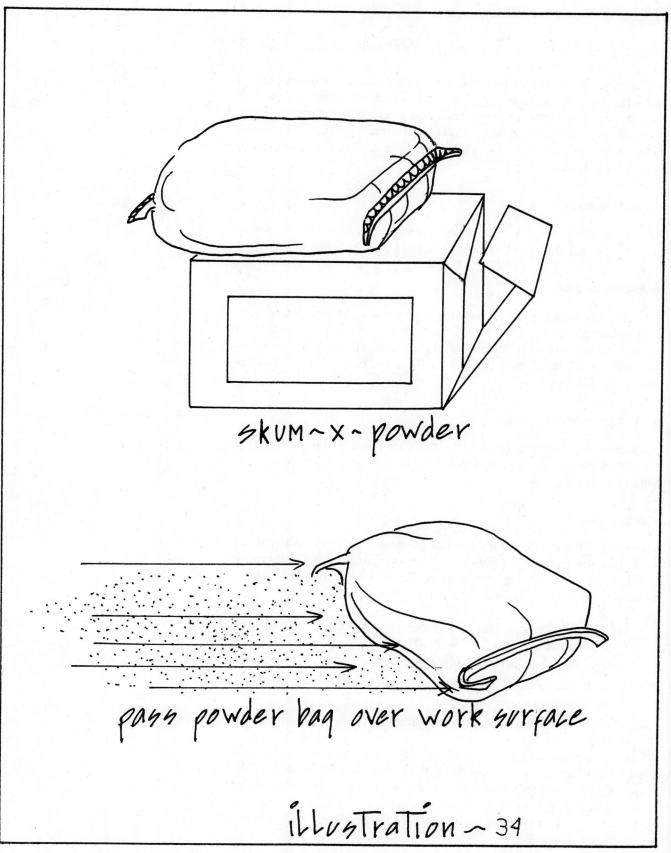

skum~x~powder

pass powder bag over work surface

illustration~34

rubbed. This, however, is a less preferable method as it is less effective and is somewhat messier.

Care should be taken when Skum X is used so that an excess film is not left on the drawing paper. This excess film will cause a pencil or pen to skip when attempting to continue drawing. Erasures must be brushed away before additional drafting commences. It is best to use Skum X at the end of a project and not while drafting is taking place so that pencil lead will not skip.

ERASING SHIELD

An erasing shield is used for covering parts of a drawing while exposing only those areas that need to be erased. (See illustration #35.)

DRAFTING BRUSH

After any erasures are made, a drafting brush should be used to remove stray erasure particles from the surface. In this manner, no additional smears can be created by the use of one's hands. However, an inexpensive, wide, paint brush may be used for the same purpose. (See illustrations #36 to #38.)

POUNCE POWDER

Pounce powder is used to prepare a drafting surface for inking. Dusting the surface with a light film of powder will provide for smoother handling of paper and ink.

SCALE RULER

Scale rulers have a tendency to confuse people at first, and without need. Just remember the following: there are four types of scale rules; architect, mechanical engineer, civil engineer, and metric. The above mentioned scale rules measure in the following way:

1. *Architect:* measures feet and inches in a smaller scale than the object is in reality, e.g., ¼″ = 1′–0″;
2. *Mechanical Engineer:* is divided into inches and parts of an inch;
3. *Civil Engineer* (also known as a decimal inch scale): measures objects in parts to the inch; and
4. *Metric:* measures an object in millimeters.

We will concern ourselves with only the architect's scale, as it is the one that is used in the drafting field. With this in mind, do not let the shape of the architect's scale ruler confuse you: it may still come in three shapes, a triangular shape or two types of beveled shapes. The shape used is just a matter of preference. (See illustrations #39 to #41.)

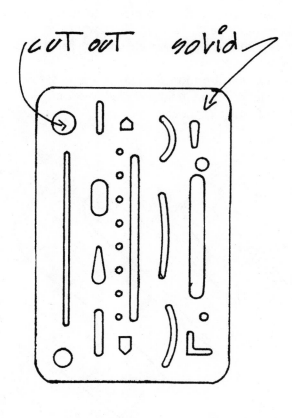

cut out   solid

erasing shield

illustration ~ 35

An architect's scale ruler may be used to indicate measurement in more than one scale. For instance, every ¼ inch indicator or mark on the rule may denote 1′–0″ of actual space. We see that on an architect's scale, 1′–0″ actual measurement can be "scaled down" also to ³⁄₁₆″, ³⁄₃₂″, or ⅜″, etc. In addition, if one wishes to blow up a detail of a drawing, one should employ a larger scale than has been used in the original drawing, e.g., 3″ = 1′–0″ instead of ¼″ = 1′–0″. (See illustration #42.)

The above are standard measurements and are understood in the drafting field to be representations of space. Otherwise, there would be no means by which we could draw a 20′–0″ wall on an eleven-inch sheet of paper.

Since ¼″ and ⅛″ scales are those that are most commonly used (¼″ for residential projects, and ⅛″ for large commercial projects), the manner in which they are read will be further explained.

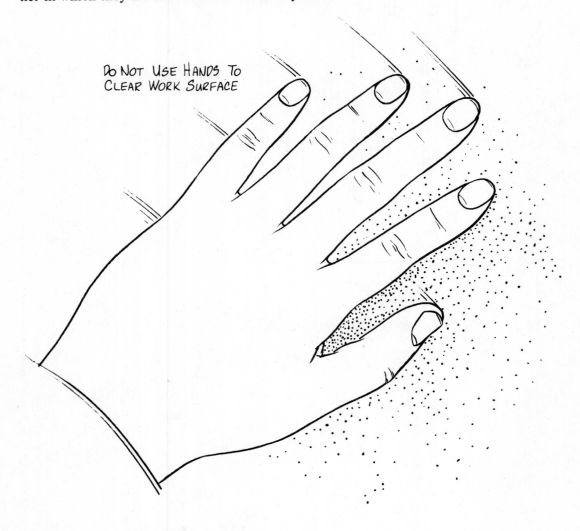

DO NOT USE HANDS TO
CLEAR WORK SURFACE

ILLUSTRATION – 36

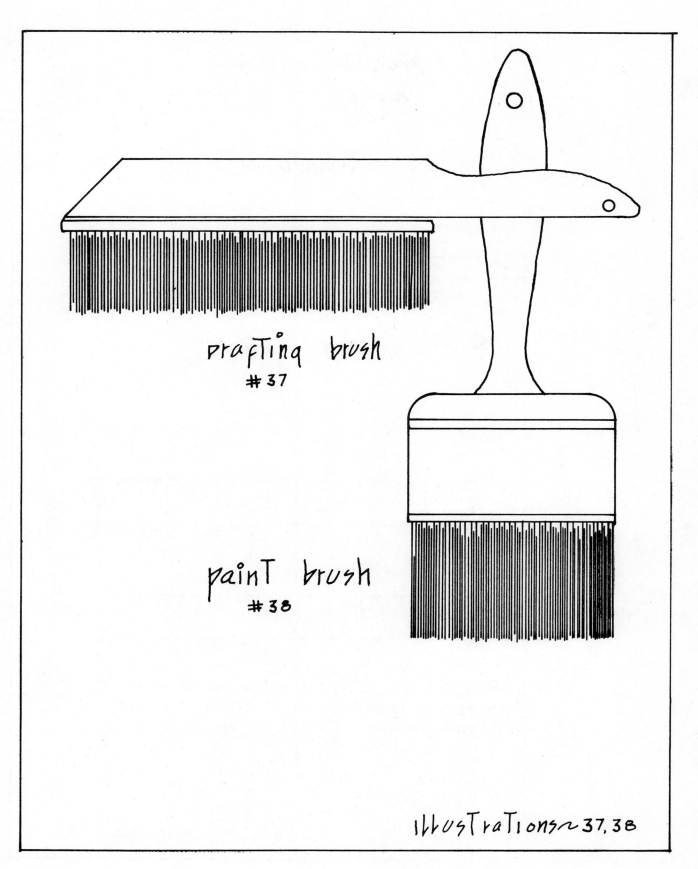

drafting brush
#37

paint brush
#38

illustrations 37, 38

Architect's Scales

#39   Triangular

#40   Flat-beveled
       4 sided

#41   Flat-beveled
       6 sided

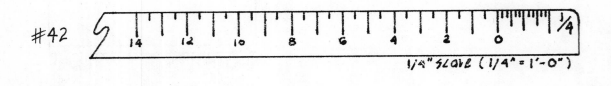

#42   1/4" scale (1/4" = 1'-0")

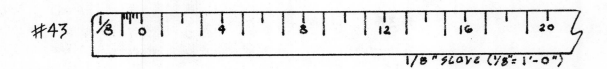

#43   1/8" scale (1/8" = 1'-0")

The above 1/4" and 1/8" scale appear on the same
face on the architect's scale. Here, they have
been separated for illustration purposes.

Illustrations 39 - 43

¼″ *Scale.* One-quarter inch scale is read from the right side of the rule to the left side. The rule is marked primarily in two-inch sections and even numbers are noted from right to left. Each number indicating an even foot measurement. The secondary demarcations directly in the center of these primary-marked sections have no numerical identification, but they are indicative of the odd number feet, i.e., 1′, 3′, 5′, 7′, etc. In addition, at the far right of the rule, where ¼″ scale has been identified, one inch in the ¼″ scale has been further divided into representative one-inch sections.

⅛″ *Scale.* One-eighth inch scale is read from left to right and is indicated in smaller size numbers than ¼″ scale. One-eighth inch scale is marked every four feet, and an even numeral representing feet has been noted. In reading ⅛″ scale (from left to right), each and every line, numbered or not, is equal to one foot. In addition, just as ¼″ scale has been further divided into inches, so has ⅛″ scale. However, ⅛″ scale in the breakdown of inches is marked every two inches rather than every one inch. (See illustrations #43 to #46.)

TRIANGLE

A triangle may come in different sizes, such as 6″, 8″, 12″, etc., and different degree angles, such as 30°, 60°, 90°, and 45°. A triangle is used to draft vertical or diagonal lines or to draft angles. In combining angles of a triangle, one may achieve a number of angles that otherwise would be difficult to draft. Usually, smaller triangles, like 6″ triangles, are used for cross-hatching, which is a rendering technique used to create shadows. The larger triangles are used for pure drafting. A triangle may also be adjustable, which permits drafting of virtually any angle. Triangles may be purchased in either metal or plastic. Plastic is preferable because of its transparency. (See illustration #47.)

FRENCH CURVE

A french curve facilitates the drafting of irregular curves. By lining up four points of contact, one may draft an irregular curve with great precision. They too come in many sizes and are usually made of plastic. (See illustration #48.)

FLEXIBLE CURVE

A flexibile curve (also called a "snake") serves the same purpose as does a french curve. However, a flexibile curve is made of rubber or pliable plastic and can be molded to any required curve, whereas a french curve is a hard piece of plastic and may have to be shifted several times to line contact points up. A flexible curve also comes in many sizes. (See illustration #49.)

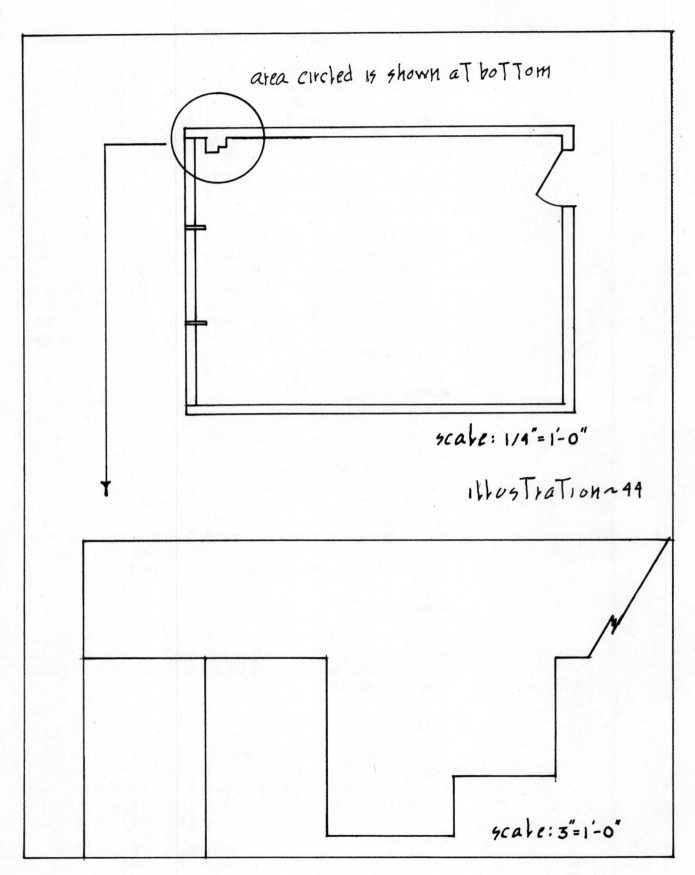

area circled is shown at bottom

scale: 1/4"=1'-0"

illustration~44

scale: 3"=1'-0"

28

Architect's Scale

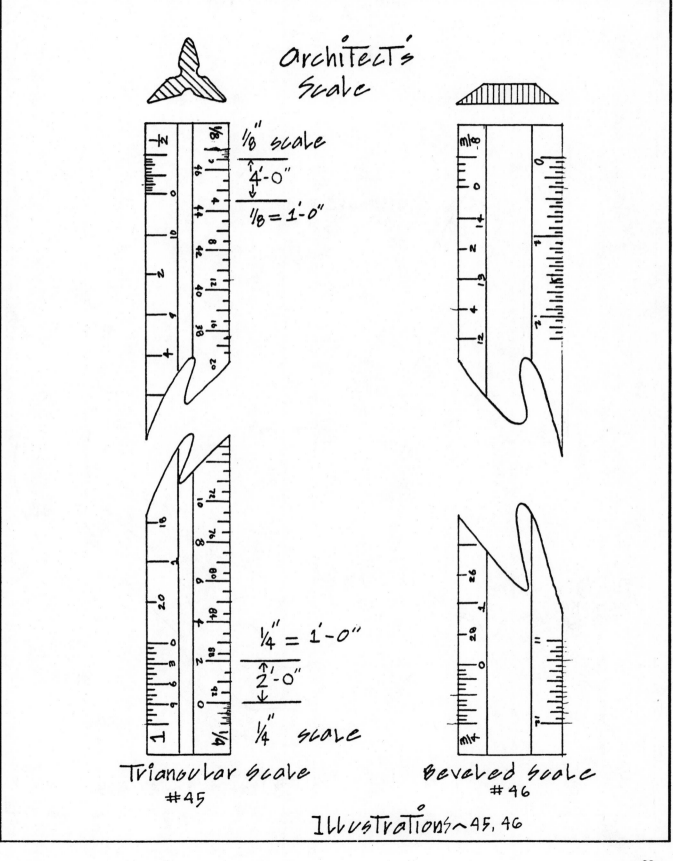

1/8" scale

↑
4'-0"
↓

1/8 = 1'-0"

1/4" = 1'-0"

↑
2'-0"
↓

1/4" scale

Triangular Scale
#45

Beveled Scale
#46

Illustrations ~ 45, 46

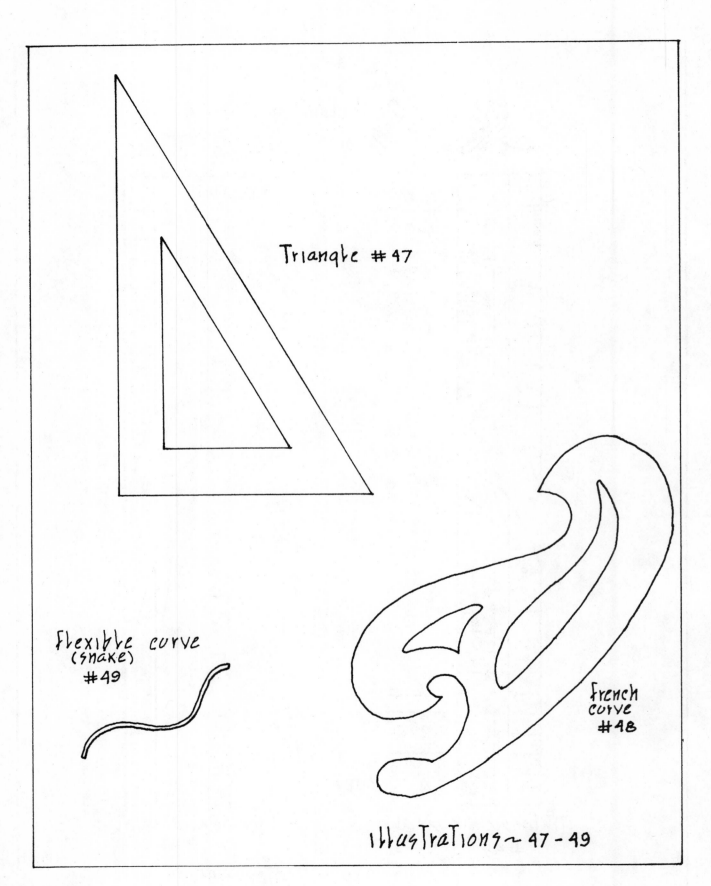

Triangle #47

Flexible curve
(snake)
#49

French
curve
#48

Illustrations ~ 47 - 49

COMPASS

A compass is the instrument used to draft parts of a circle, as well as some common curves. One's drafting pencil is inserted in one end of the compass while the other end of the compass is used to maintain a stationary point. Compasses are adjustable to permit the size of the arc (portion of the circle) or curve to be variable. They come in three types, each type dependent upon the size. Usually, the larger compasses (about 6″ in length) are called bow compasses, the medium size are called friction compasses, and the smallest size are called center wheel adjusting compasses. When drafting a straight line and adjoining it to a curved line, always draft the curved line first to avoid mismatched points. (See illustration #50.)

PROTRACTORS

Protractors are used to measure angles in degrees. This degree measurement assists a contractor in constructing more accurately a specified design element. Protractors may be semi-circular or circular. (See illustration #51.)

DIVIDERS

Dividers are used to divide a given line into equal segments or to indicate measurement of a certain length line that must be repeated several times. They come in basically two sizes: either 6″ or 4″ in length. They can save a great deal of time when used properly. (See illustration #52.)

BURNISH PLATES

A burnish plate is an underlay device used to express texture in a drawing. It is placed under a sheet of paper, and a pencil is used as a burnishing device by spreading lead over the surface. Burnish plates have raised and recessed surfaces. The lead only touches the raised surfaces and an impression is left by the lead. These, however, are rarely used in the drafting field due to newer innovations mentioned. In all probability, they will be difficult to locate and purchase.

GENERAL UNDERLAY

Underlays are placed beneath a sheet of paper, and the desired outline or shape is traced on the paper. Underlays may be purchased in a variety of shapes such as trees, bricks, stone, and so on. Again, these are rarely used in drafting because of newer innovations, and would be difficult to locate and purchase.

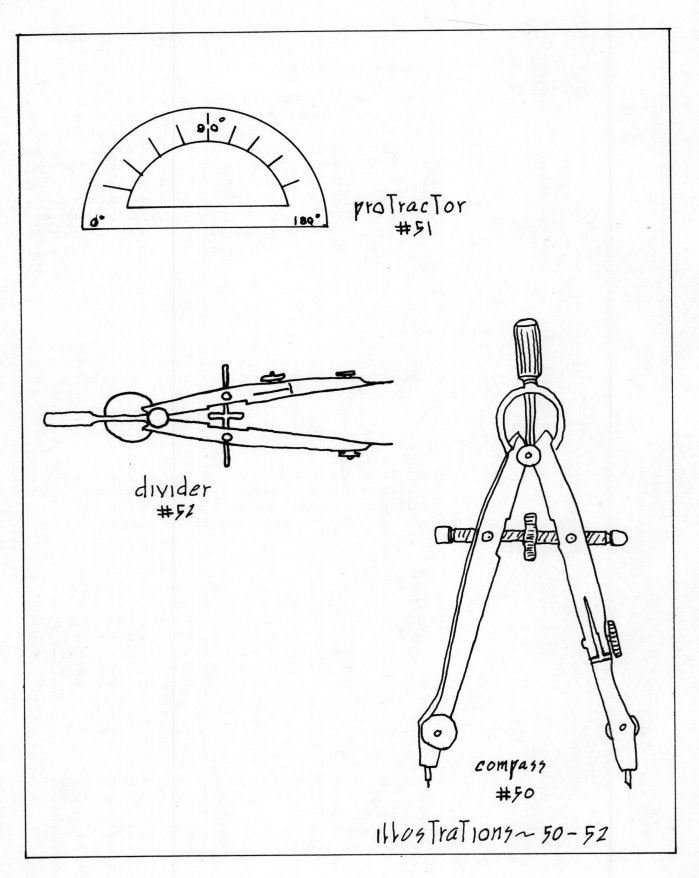

proTracTor
#51

divider
#52

compass
#50

illusTraTions ~ 50-52

GENERAL OVERLAY

An overlay is a transparent sheet that permits the viewer to see a drawing with embellishments, such as trees, or varied rendering techniques, such as stone, brick, or trees. Overlays come in two types. The first type peels off a transparent sheet and sticks directly onto the drawing. They are similar in usage to color forms, which we used as children. The second type is not removable from the plastic sheet, and it is only a temporary method of placing an overlay on the drawing. (See illustration #53.)

RUBBER STAMP

Rubber stamps are a time saving method of indicating trees or architectural symbols. They should be restricted in use so as not to give one's drawing a manufactured or prefabricated look due to overly repeated shapes of the same character. (See illustration #54.)

LETTERING GUIDES

A lettering guide gives one's printing a uniform look. It is used as a template and can assist your drawing in looking professional if your own printing is not consistent. (See illustration #55.) However, it is much more preferable that one learn to print uniformly without the aid of a lettering guide.

TRANSFER LETTERS AND SYMBOLS

Transfer letters or symbols are also considered an overlay device of the permanent type. Letters or symbols just peel off a plastic backing and adhere to the drafting paper. (See illustration #56.)

TEMPLATES

Templates are plastic shields out of which special shapes have been cut. For example, there are circle templates, triangle templates, office furniture templates, and residential furniture templates, bathroom fixture templates, etc. These cutouts are in scales, for example, $\frac{1}{4}'' = 1'-0''$; $\frac{1}{8}'' = 1'-0''$, and so on. (See illustrations #57 and #58; both are in $\frac{1}{4}''$ scale.)

Needless to say, templates can save limitless amounts of time. The novice, however, should not become dependent on templates as a means of drawing shapes until he or she is capable of drafting the same shape to scale without its use.

Any of these supplies should be available at your local major art supply store.

G Manufactured and printed in U.S.A.
GRAPHIC PRODUCTS CORPORATION

FORMATT No. 6633

Illustration 53

34

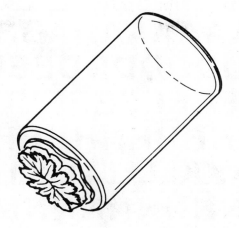

Handstamp #54

abcdefghijklm:nopqrstuvwxyz

Berol RapiDesign.   R-946   SLANTED LETTERING GUIDE $\frac{3}{16}$ IN.

1234567°°90&()

ABCDEFGHIJK:LMNO.PRSTUVWXYZ

illustration # 55

AAAAAAAAAAAAAAAABBBBBC;;
CCCCCCDDDDDDDDDEEEEEEE;
EEEEEEEEEEEEEEEEEEEFFFFF;
GGGGGGGHHHHHHHHHHHIIIIIII
IIIIIIIJJJKKKKKLLLLLLLLLLLLLLM;
MMMMMMNNNNNNNNNNNNNN;
NNNNNOOOOOOOOOOOOOOOPP
PPPPPQQQQRRRRRRRRRRRS
SSSSSSSSSSSSSSTTTTTTTTTT;
TTTTTTTUUUUUUUUUUUUUUUV;;
VVVVVWWWWWWWXXYYYYYZZZZZ
äàaaaaaaaaaabbbbcccccccd;;
dddddddèééeeeeeeeeeeeeeeee
eeeeffffgggggggghhhhhhhhhhhiii;
iiiiiiiiiijjkkkkllllllllllmmmmmmn;
nnnnnnnnnnnnnnnöoooooooooo
oppppppppqqqrrrrrrrrrrssssss
sssssssssttttttttttttttttttüüûuuuuu;
uuuuvvvvwwwwwwwxxyyyyzzzz;
1111222233334444555566667;
77888899990000&?!ß£$()✿⅝;;

illustration #56

FURNITURE TEMPLATE

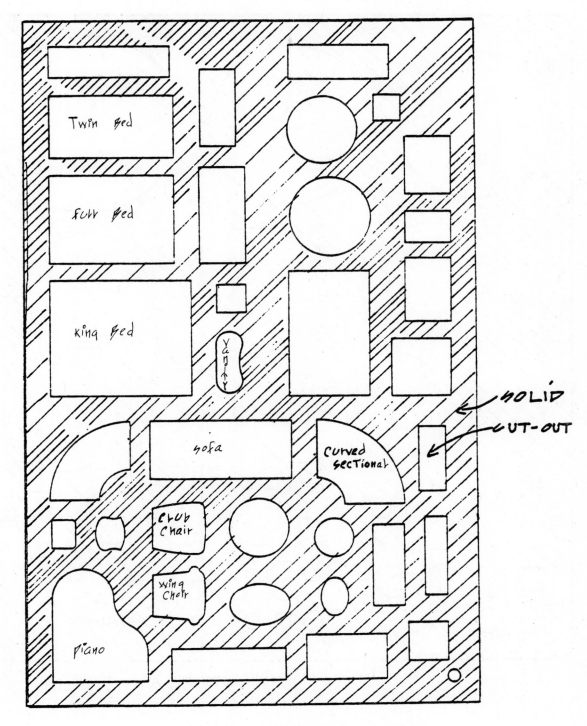

Illustration~57

CIRCLE TEMPLATE

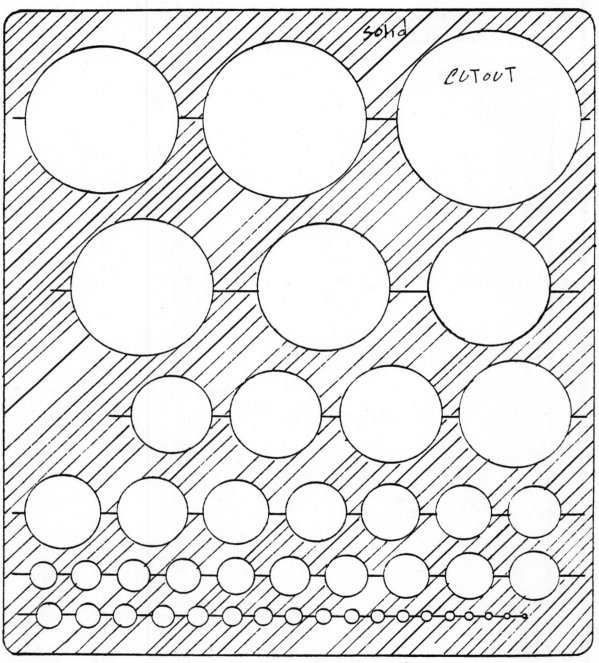

Illustration ~ 58

REVIEW OF IMPORTANT POINTS

1. Use a sturdy, level drafting surface.

2. Tape drafting paper down in a secure and tight manner, placing drafting tape on the diagonal axes at each corner.

3. Use the proper tool for each task:
   a. T-Square or parallel rule for drafting horizontal lines or supporting other tools;
   b. Triangle for drafting vertical or diagonal lines, or for cross-hatching;
   c. Sketch or rough draft paper for sketches or rough drafts;
   d. Medium grade paper for preliminary drawings and general layout work;
   e. Quality grade (Vellum) paper for finished drawings;
   f. Films for inking or permanent drawings;
   g. HB lead for primary drawings or sketches;
   h. F & H leads for finished drawings, layouts, and lettering;
   i. 2H lead for finished drawings;
   j. 4H lead, etc. for precise layouts;
   k. Lead pencil, mechanical pencil, or leadholder for general drafting;
   l. Technical pen for presentation and permanent drawings;
   m. Softest erasers possible;
   n. Erasing shield to protect the area of drawing not to be erased;
   o. Skum X to camouflage smudges;
   p. Mechanical sharpener or sandblock for sharpening lead;
   q. Drafting brush to brush away erasures, do not use hands;
   r. Pounce powder to prepare paper for inking;
   s. Architect's scale ruler: $\frac{1}{4}'' = 1'\text{-}0''$ for most projects, $\frac{1}{8}'' = 1'\text{-}0''$ scale for larger projects, $1'' = 1'\text{-}0''$ or larger for detail drawings;
   t. French curve, snake (flexible curve) for irregular curves;
   u. Compass for circles or arcs;
   v. Protractor for measuring degree of angle; and
   w. Templates to provide outline of furniture, or architectural elements to scale;

4. When using a T-square or parallel rule to tear drafting paper, use a swift motion.

5. Lead should be constantly needle-sharp (sharpen often, every three to five lines drawn).

6. Rotate leadholder or mechanical pencil between fingers when drafting. Pencil is kept in motion to assist in keeping pencil point sharp.

7. Pull lead across the paper, do not push it across.

8. Keep drafting hand slightly angled and elevated to help eliminate smudging.

QUIZ QUESTIONS

1. A _____ is to horizontal lines as _____ is to vertical lines.

2. One may use either a _____ or a _____ to draft horizontal parallel lines.

3. Name the grade of paper that can be used for rough drafts or sketches and the grade that can be used for overlays.

4. Name the grade or type of paper that can be used for finished drawings.

5. Drafting tape is more or less sticky than masking tape?

6. In what direction should drafting tape be applied to a sheet of paper?

7. What are the differences between a lead pencil, mechanical pencil, and lead holder?

8. How can one prolong the sharpness of a point of lead pencil, mechanical pencil, and lead holder?

9. Lines are drawn from _____ to _____ for left-handers or from _____ to _____ for right-handers.

10. Name the various types of leads and their characteristics and uses.

11. Name the various types of erasers and their uses.

12. Name three types of sharpeners.

13. When erasing, what can be used to protect the area of the drawing not to be erased?

14. What instrument is used to brush erasures away?

15. What instrument camouflages smudges?

16. What are the most common scales used on an architect's scale ruler?

17. What are the differences between an architect's scale and a civil and mechanical engineer's scale ruler?

18. Name four types of templates and give their respective uses.

EXERCISES

1. Practice cutting tracing paper with your T-square.

2. Practice rotating your lead pencil, mechanical pencil, or lead holder while drawing a horizontal or vertical line.

3. Experiment with various types of leads to ascertain how each reacts to your hand pressure.

4. Practice drafting lines of differing widths with various pen points.

5. Looking at your scale rules, make a mental picture of equal measurements on different scales.

2
BASIC CONCEPTS OF GRAPHIC REPRESENTATION FOR THE NOVICE DRAFTSPERSON

Before discussing the many types of plans that a draftsperson will create, a somewhat in-depth knowledge of the various graphic methods employed by the draftsperson to represent the architectural or decorative elements of any particular plan must be made available. In addition, graphic methods of properly indicating measurements, notations or explanations, degree of importance of any architectural or decorative element, and any indication of additional information concerning the elements of a drawing must also be made available. This chapter is devoted to each of the above aspects by isolating each aspect; in essence, it removes them from the context of a drawing or plan and gives an explanation to each.

ARCHITECTURAL LETTERING

As stated in the foreword, each person has his or her own style of drafting. The same principle holds true for any printing that needs be done either on the drawing proper or in the legend and title boxes. (Legend and title boxes will be explained in Chapter 3.) One may have a very simple method of printing or a somewhat more embellished style. Either is correct. (See illustration #59.)

Basically, there are three types of printing—forward slant, vertical (no slant), and back slant. It is important that one use the style that is most comfortable for oneself, thereby facilitating printing. Not only will one be able to print faster, but in addition, printing will not seem strained and will therefore be easier to read. (See illustrations #60 to #62.)

When printing on the drawing proper, or in the legend or title box, letter size should be approximately ⅛″ to ³⁄₁₆″ high. However, if one wishes to bring special attention to a written notation, one may do so by using italics, or full ¼″ high lettering. In any case, ¹⁄₁₆″ to ⅛″ spacing should be maintained between lines of print. All letters should be upper case to help give printing style a sense of uniformity.

SPECIFIC TECHNIQUES OF LETTERING

- Always use horizontal guidelines to help control uniformity of lettering height (see section on line weights).

- All vertical strokes of lettering are drafted in a downward motion and are drafted first.
- All horizontal strokes of lettering are drafted from left to right (for right-handers, and the reverse for left-handers) and are drafted second.
- All curved strokes are drafted from the top to the bottom and are drafted third.
- Spacing between letters should not, in actuality, be equidistant, but instead, should just appear equidistant.
- Spacing between words should be equal to the height of the lettering used.
- Spacing between sentences should be maintained at twice the height of the lettering used.

(See illustration #63.)

printing

A B C D E F

block

A B C D E F

serif

Illustration ~ 59

FOWARD SLANT #60

VERTICAL #61

BACK SLANT #62

ILLUSTRATIONS — 60-62

leTTering

A B C D E F G H I J K L
M N O P Q R S T U V
W X Y Z

illusTraTion ~ 63

LINE WEIGHTS

Line weights (or the pressure one exerts on a pencil, resulting in light or dark lines) are used for specific purposes. Basically, there are three varying degrees of line weight used for drawing plans. The eye is directed initially to darker lines and lastly to very light lines. Thus, varying line weights are employed as follows:

1. *Guideline Weights.* Lines that should be barely visible. They are drawn as blocking or mere "guides" at the top and bottom of labeling to create uniform lettering and numbering. In addition, some preliminary drafts use guide lines to facilitate inking at a later time. Guidelines should be so light that one must almost strain ones eyes to see them. In this way, guidelines will not reproduce when blueprinted.

2. *Medium Weights.* Lines that are drawn for representation of items of *secondary* importance in the drawing in addition to note and measurement indication; and

3. *Heavy Weights.* Lines that are drawn for representation of items of primary importance in the drawing. Since we are attracted initially to darker lines, these will be observed first.
   (See illustrations #64 to #66.)

Decisions as to what elements are of primary or secondary importance in a drawing are clearly that of the draftsperson or designer. However, they are usually determined, as a standard, by the title or in other words, the type of drawing. A combination of all three weights should be employed in a drawing in order to insure visual attractiveness. The contrast of line weights not only creates interest, but simplifies a drawing's reading. For instance, in a furniture plan, as its title suggests, the furniture is of the utmost importance. Therefore, a heavy line weight would be used for furniture. The architectural outline of the plan, measurement indication, and labeling or notations are of secondary importance in such a drawing, and therefore a medium weight would be used for that portion of the drawing.

When using a leadholder or mechanical pencil, line weight is determined by the pressure one exerts on the pencil and the type of lead used when drawing. When using a drafting pen, line weight results from the use of varying thicknesses of interchangeable pen points. (See Chapter 1, *Tools of the Trade.*)

MEASUREMENT INDICATION AND SPECIAL NOTATIONS

When drafting measurements of any sort, one's main concern is having both the beginning and end points of that measurement easily found and read. It must, in other words, have clearly recognizable points of reference. In addition, being consistent in one's style of measurement notation is equally important. Ambiguities and inconsistencies permit

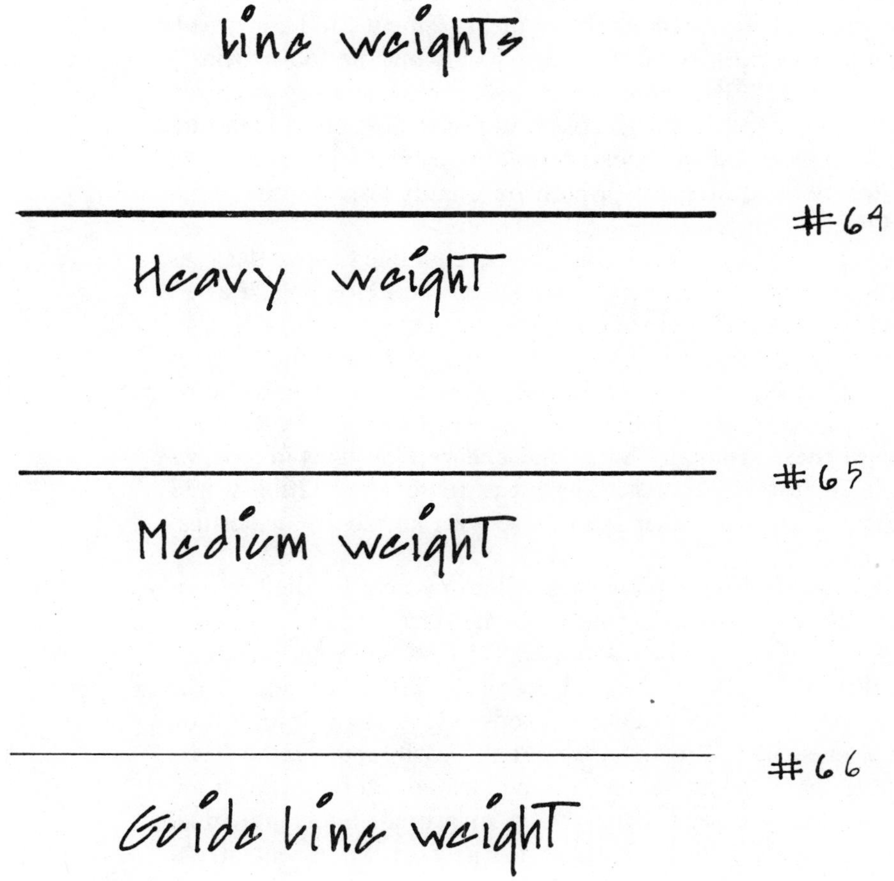

line weights

_____ #64
Heavy weight

_____ #65
Medium weight

_____ #66
Guide line weight

ILLUsTraTions 64-66

the chance of misinterpretation, thereby resulting in a problem for those who will be utilizing the measurements and those concerned with the areas that these measurements represent.

The following are standard methods of representing measurement information. In addition to clearly defined beginning and end points, the numerical representation should always be expressed in feet and inches, and when possible, displayed in a break in the center of the line that shows the span of measurement. (See illustrations #67 to #73.)

By always expressing a measurement in feet and inches (i.e., 0'–2", 1'–7", 12'–0", etc.) it is less likely that a measurement can be misread. One must also be certain, however, that when using the foot mark (') and the inch mark (") that these too are easily read and separated by a hyphen (-). An exception to this rule is in the indication of furniture dimensions, kitchen appliances, counters, and cabinets in addition to certain construction requirements, which are usually displayed in inches only (e.g., 53", 126", etc.).

When areas to be measured intersect, it is acceptable to have the lines indicating the span of measurement cross each other. However, if at all possible, it is preferable if they do not cross. An excessive crossing over of measurement lines should be avoided in order to dimish the chances of confusion and misinterpretation. To the same end, one should choose the method of measurement indication that will be most clearly read. For instance, if there are many horizontal and vertical lines in a drawing, the better method of representing measurement indication would be any of these illustrated other than those with horizontal or vertical end strips.

To facilitate the reading of plans, note all dimensions on the horizontal plane. In this way, a drawing need not be turned in order to read measurements. (See illustrations #74 and #75.) Measurements may be indicated either in the body of the drawing or on the outside of the drawing. One may use either or both methods as long as the information given is being presented in its most logical and easily interpreted form. It is preferable that a measurement not be located too far away from, or too close to, that element that is being measured. Each individual must make his or her own decision as to the best placement of the measurement.

Sometimes, it may be preferable to give a measurement from the center of an element to the center of another element, such as in the case of a reflected ceiling plan or electrical plan. However, if this method is used, it should be used consistently throughout that particular drawing. Here again, consistency is one of the keys of success. (See Chapter 4, the sections on *Reflected Ceiling Plan* and *Electrical Plan.*) Measurements should always be exact; however, as a precaution, it is advisable that the blanket statement "all measurements to be verified by trade in field" be placed in the legend box on each plan.

If a special notation need be made on the plan or if insufficient room is available to permit measurement indication in its proper location, an arrow may be drawn (called a leader) toward the graphic representa-

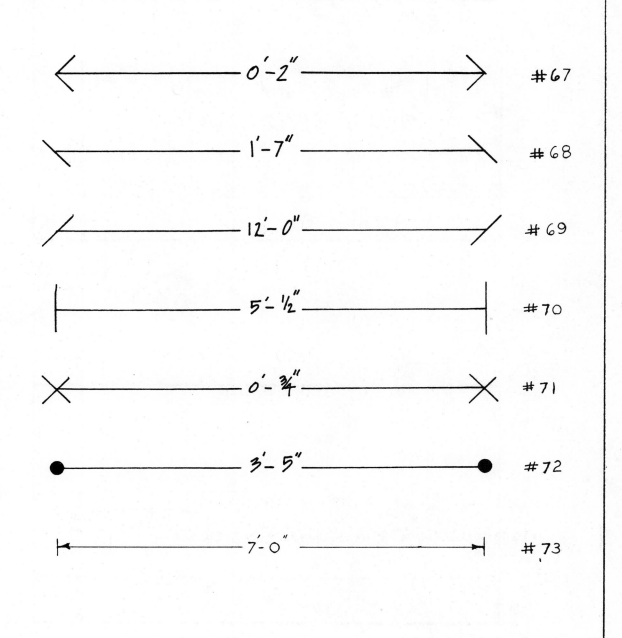

ILLUSTRATIONS~ 67-73

49

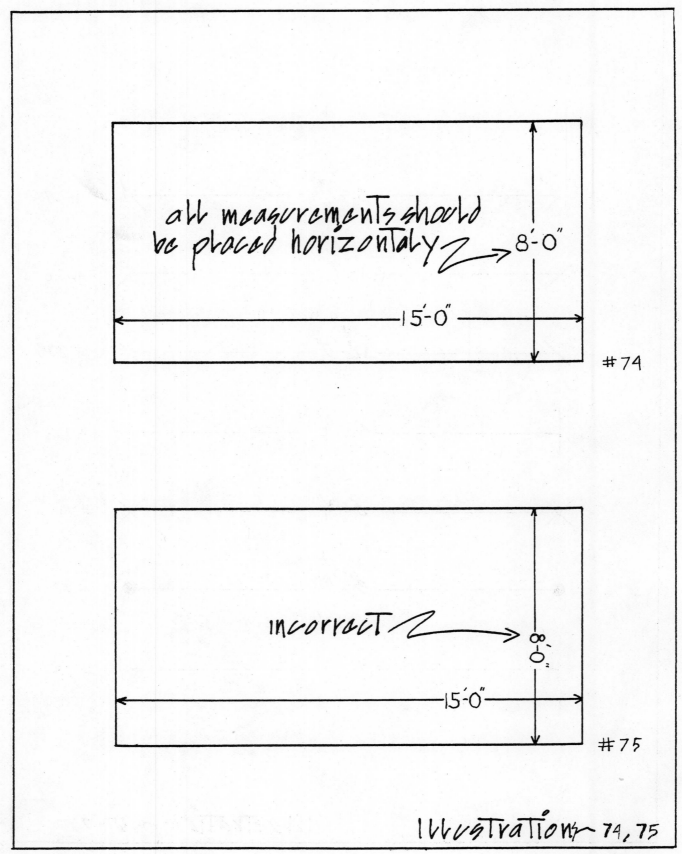

all measurements should
be placed horizontaly → 8'-0"

15'-0"

#74

incorrect → 8'-0"

15'-0"

#75

Illustrations 74, 75

tion and the note, detail, or measurement written next to the leader. The shape of the leader should be dictated by the linear structure of the drawing. For example, if the drawing is composed of a great many curves (curvilinear), then a straight leader should be employed. If the drawing is composed of a great many horizontal and vertical lines (rectilinear), a curved leader should be used. (See illustrations #76 to #79.)

Breaklines

To indicate that an element is not being graphically displayed in its entirety, such as additional steps, remaining wall lengths, etc., the following symbol is used at the break of the element: ⌒. (See illustration #80.) In essence, what this symbol is saying is that this particular element does continue, but for purposes of clarity in reading the plan where continuation of this element is not essential to the plan, the element is being stopped at this point.

WALLS

In reality, by construction, a wall has two sides and therefore must be represented by two lines that are parallel to each other. One line represents one side of the wall and the other line represents the other side. In representing a wall as such, it permits space for additional representation of other architectural elements, such as doors and windows. (A door or window, in reality, is placed between the two parallel sides of a wall.) Walls may vary in width due to their particular construction and would therefore vary in width when drafted in scale. In addition, the composition or physical make-up of a wall must be indicated. This is graphically represented in a simplified form by the use of pouchering. (See illustrations #81 to #85.)

POUCHERING

Pouchering is the term used to describe the way in which we graphically represent the material composition of any constructed element, for example, a plaster or wall-board wall, a wood or steel column, a stone or concrete floor, and so on. (See illustrations #86 to #101. These indicate some standard pouchering techniques that are universally understood in the construction field.)

DOORS AND WINDOWS

Since there are many types of doors and windows available on the market today, one must familiarize oneself with the various ways in which a specific type of door or window may be presented on a floor plan.

The following are typical representations of types of doors and windows, giving both a floor plan (bird's-eye-view) and a frontal view with slight perspective. These representations may differ slightly according

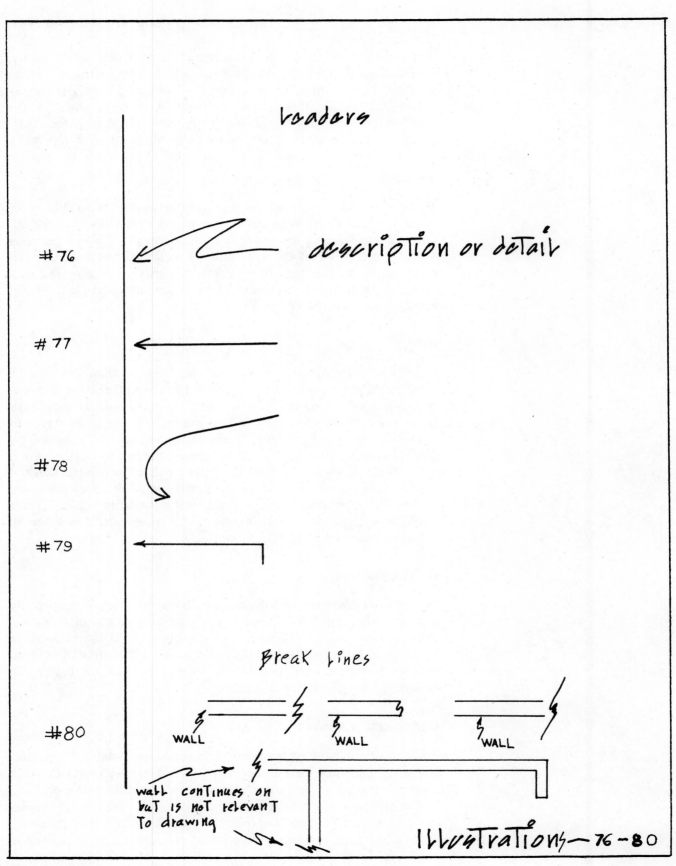

leaders

#76   description or detail

#77

#78

#79

Break Lines

WALL   WALL   WALL

#80

wall continues on
but is not relevant
to drawing

ILLUSTRATIONS — 76-80

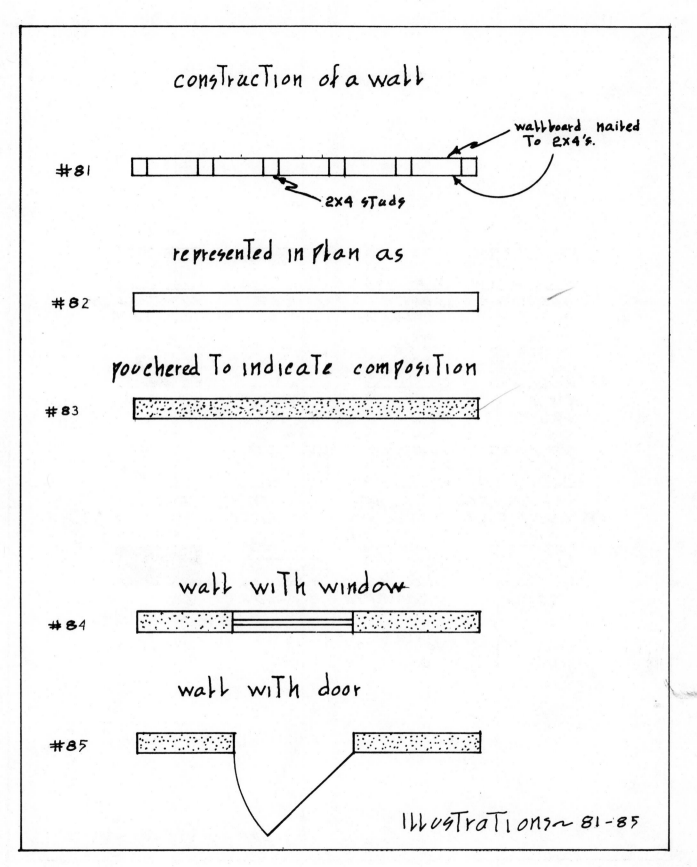

construction of a wall

wallboard nailed
To 2x4's.

#81

2x4 studs

represented in plan as

#82

youchered To indicate composition

#83

wall with window

#84

wall with door

#85

ILLUSTRATIONS 81-85

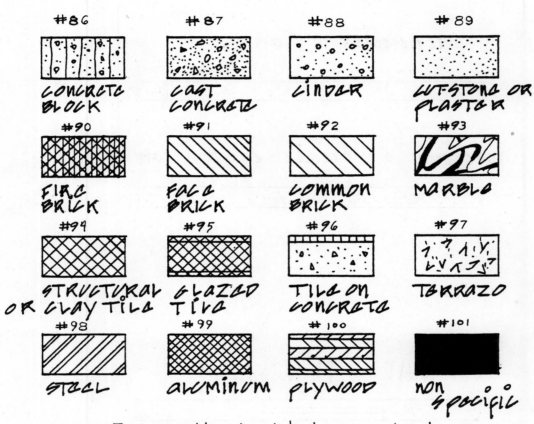

#86 CONCRETE BLOCK
#87 CAST CONCRETE
#88 CINDER
#89 CUTSTONE OR PLASTER

#90 FIRE BRICK
#91 FACE BRICK
#92 COMMON BRICK
#93 MARBLE

#94 STRUCTURAL OR CLAY TILE
#95 GLAZED TILE
#96 TILE ON CONCRETE
#97 TERRAZO

#98 STEEL
#99 ALUMINUM
#100 PLYWOOD
#101 non specific

The entire wall should be pouchered with one of the above Techniques.

ILLUSTRATIONS ~ 86-101

54

to one's own preference for a particular drafting style. A door may be represented by a single line or double line. (See illustrations #102 to #111.)

Note: Remember, a door must be able to fit in its opening; therefore, be sure that the opening and the door widths are of equal lengths. For consistency, and therefore for aesthetic reasons, all door swings should be drafted in an open position and at the same angle, usually 60° to 90°. In addition, be sure that the door is drafted at the point along the wall where hinging will actually take place. (See illustrations #102 to #111 again.)

Windows

Notation for windows is somewhat more standard than that of doors. There are, however, some variations, such as inclusion of a window sill, that will influence the way in which one represents a window in drafting. (See illustrations #112 to #121.)

Again, individual style permits some flexibility to what is indicated.

Drafting all vertical mullions (the architectural structure that holds together panes of glass) is also important and must therefore be included in one's plans.

Above-Eye-Level Elements and Hidden Elements

For elements above eye-level, such as beams, archways, shelves in kitchen cabinets, or closets, and lofts, a line made by a series of uniform-length dashes is used (- - - - - -). These dashes should form the shape or outline of the particular element. In the case of shelves, representation is made by a single line of dashes indicating only the front of the shelf. If two or more elements that are above eye-level cross each other, it would be advisable to use two different size dashes, one for each element. One must remember that the dashes simply indicate that something is happening above eye-level. They do not explain what is happening. Therefore, one must use a leader arrow pointing to these dashes and give an explanation of their representation, for example, archway, skylight, etc. (See *Existing Conditions Plan* in Chapter 3.) If something is hidden, such as a chair seat under a table, it would be best to use a dot and dash for the hidden portion of the element. (See illustration #122; see Chapter III, the sections on *Existing Conditions Plan* and *Furniture Plan*.)

STEPS AND PLATFORMS

Each step may be indicated by a single line with an arrow indicating how many steps are involved and the height of the riser of each step. It is also important to indicate with this leader arrow the direction (up or down) the steps go while facing them. The tread depth, as well as the width of the steps, must be noted by use of a leader arrow and mea-

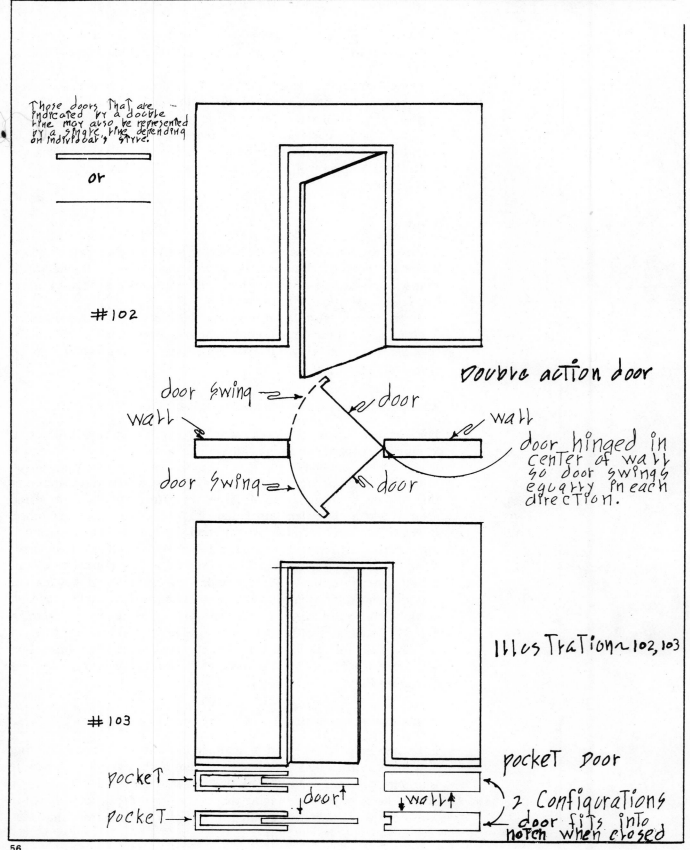

Those doors that are
indicated by a double
line may also be represented
by a single line depending
on individual's style.

or

#102

Double action door

door swing

wall

door

door swing

door

wall

door hinged in
center of wall
so door swings
equally in each
direction.

Illustration 102, 103

#103

pocket

pocket

door

wall

Pocket Door

2 Configurations
door fits into
notch when closed

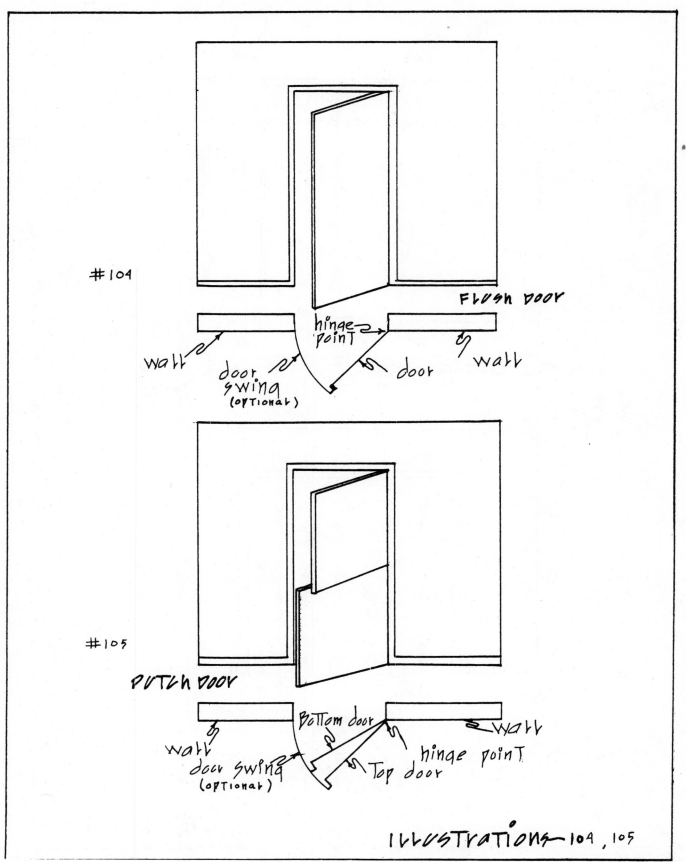

#104

FLUSH DOOR

hinge point

wall

door swing
(optional)

door

wall

#105

PUTCh DOOR

Bottom door

wall

wall

door swing
(optional)

Top door

hinge point

ILLUSTrations 104, 105

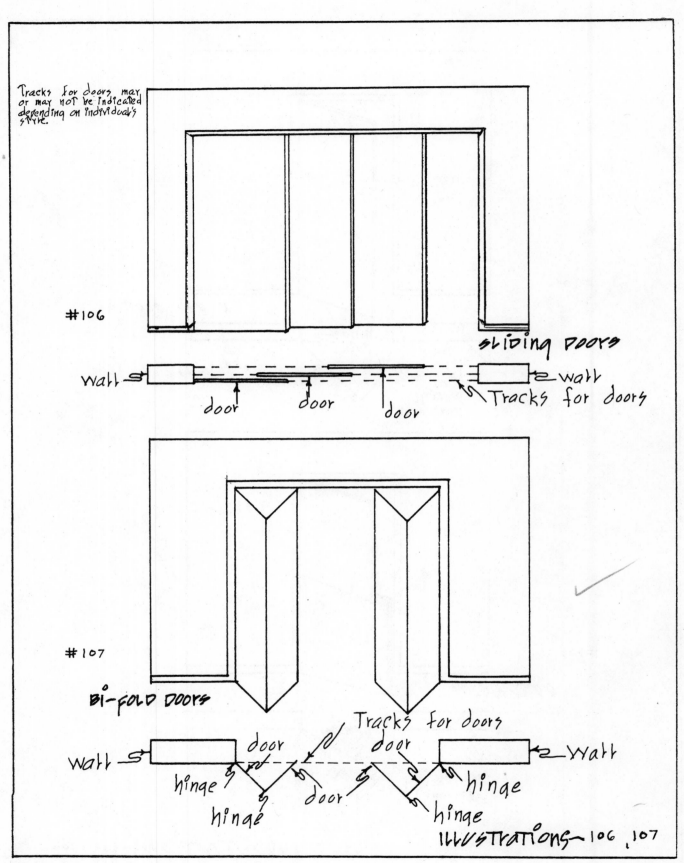

Tracks for doors may
or may not be indicated
depending on individual's
style.

#106

SLIDING DOORS

Wall · door · door · door · Wall Tracks for doors

#107

BI-FOLD DOORS

Tracks for doors

Wall · door · door · door · Wall

hinge · hinge · hinge · hinge

ILLUSTRATIONS - 106, 107

ACCORDION DOORS

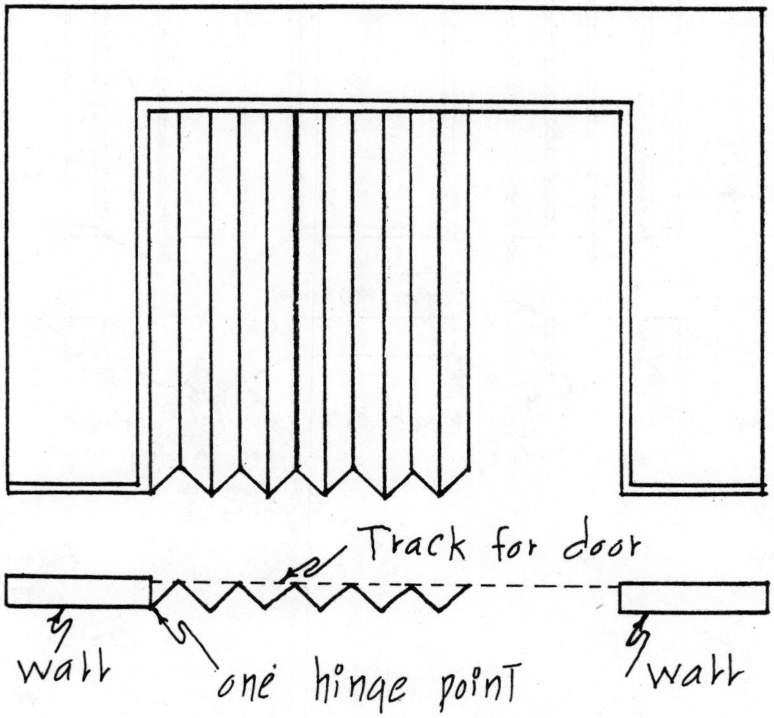

#108

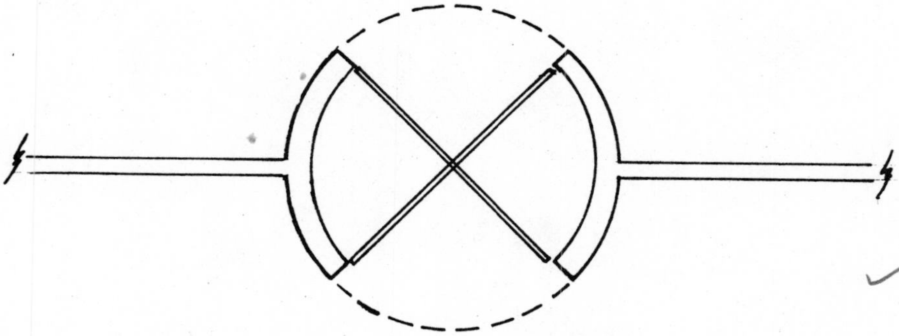

Track for door

Wall

one hinge point

Wall

#109

revolving door

ILLUSTRATIONS 108, 109

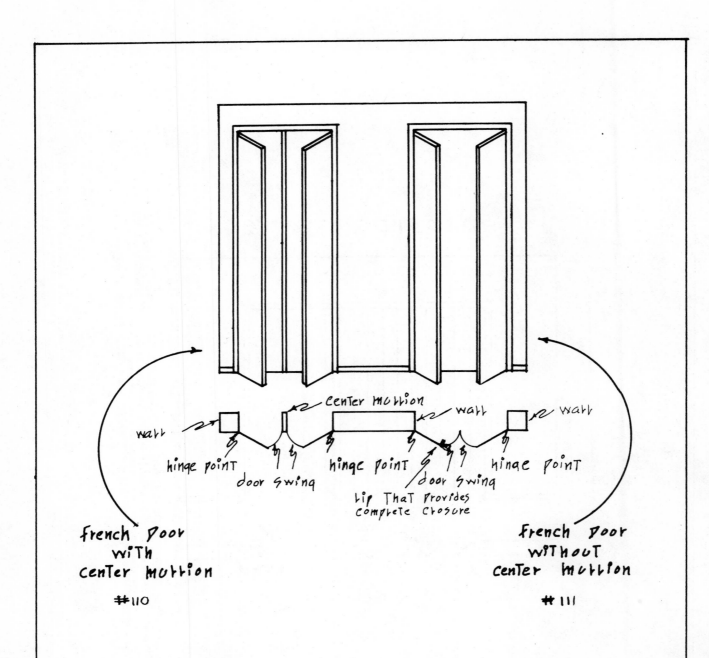

center mullion

wall

hinge point

door swing

hinge point

wall

wall

hinge point

door swing

lip that provides
complete closure

french Door
with
center Mullion

#110

french Door
without
center Mullion

#111

ILLUSTRATIONS ~ 110, 111

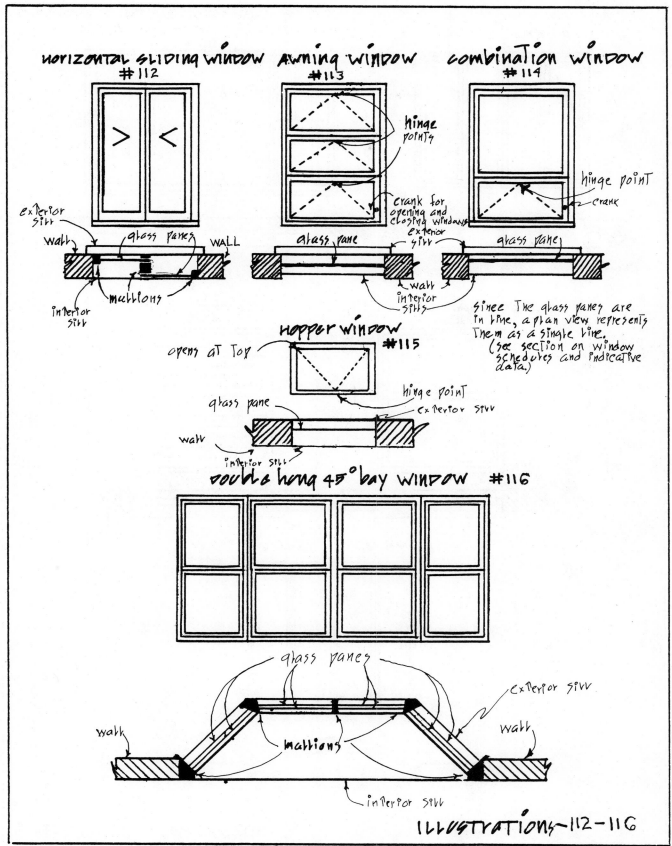

HORIZONTAL SLIDING WINDOW #112

exterior sill
Wall
glass panes
WALL
mullions
interior sill

AWNING WINDOW #113

hinge points
crank for opening and closing windows
glass pane
exterior sill
Wall
interior sills

COMBINATION WINDOW #114

hinge point
crank
glass pane

since the glass panes are in line, a plan view represents them as a single line.
(see section on window schedules and indicative data.)

HOPPER WINDOW #115

opens at top
glass pane
wall
interior sill
hinge point
exterior sill

DOUBLE HUNG 45° BAY WINDOW #116

glass panes
exterior sill
wall
wall
mullions
interior sill

ILLUSTRATIONS - 112 - 116

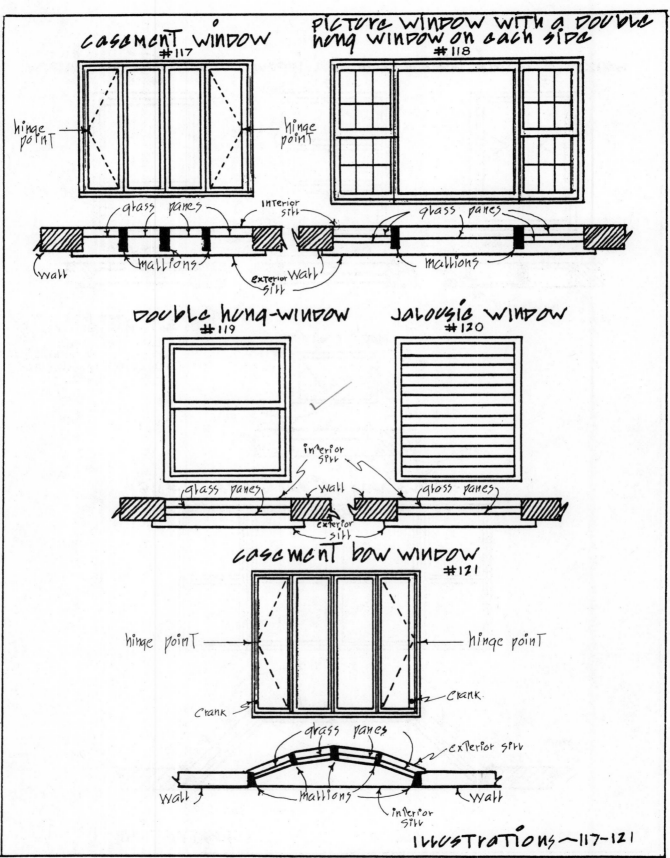

CASEMENT WINDOW
#117

PICTURE WINDOW WITH a DOUBLE HUNG WINDOW ON each side
#118

hinge point

hinge point

glass panes

interior sill

glass panes

wall

mullions

exterior sill   wall

mullions

DOUBLE HUNG WINDOW
#119

JALOUSIE WINDOW
#120

interior sill

glass panes

wall

glass panes

exterior sill

CASEMENT BOW WINDOW
#121

hinge point

hinge point

crank

crank

glass panes

exterior sill

wall

mullions

wall

interior sill

ILLUSTRATIONS ~ 117-121

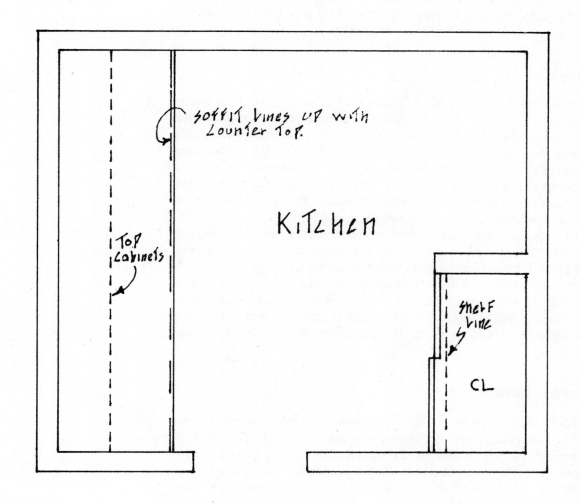

Illustration #122

surement. (See illustrations #123 to #128 and Chapter 3, the section on *Construction Plan*.)

FIREPLACES

There are two standard methods of representing a fireplace. The detailed method is usually preferred for construction drawings and the simplified version for furniture plans to be given to a client, since the construction contractor needs more detailed information than does the client. (See illustrations #129 and #130.)

SCHEDULES

A schedule is a format used to further specify in full detail particular elements of a project. For example, one may have a Door Schedule, Window Schedule, Hardware Schedule, or Finish Schedule. A schedule supplies the contractor with all the information available concerning that particular type of architectural element. This information is of the utmost importance to the contractor if he is to do his job properly, that is, without misinterpretation or error. The more information with which he is supplied, the easier his job will be.

The important points to remember are to present the schedule in a logical format, give all available data concerning each element, and key each graphic element on the plan to the schedule. Use keys in the schedule to relate it to the information given. (See illustration #131.) The following are standard formats for two of the above-mentioned schedules. Usually, schedules are presented on the plan to which they are most logically related, e.g., construction. (See illustrations #132 and #133.)

INDICATIVE DATA

Indicative data includes such information as room number, ceiling height (if necessary), and occupancy (applicable in cases of a commercial project). Means of entry and compass direction would also be considered indicative data. One should always indicate which direction is North with an arrow or other symbol, along with the letter "N" (for North). This helps establish a point of reference when needed. In addition, the main entrance location should also be noted, usually by the symbol ♠ placed at the door and the word "enter." There is no standard format to present this information that would be considered more correct than another. Therefore, one should devise a style with which one is comfortable.

It is important to show in the legend what format is being used and then use that format in a logical location on the plans most appropriate. (See illustration #134.)

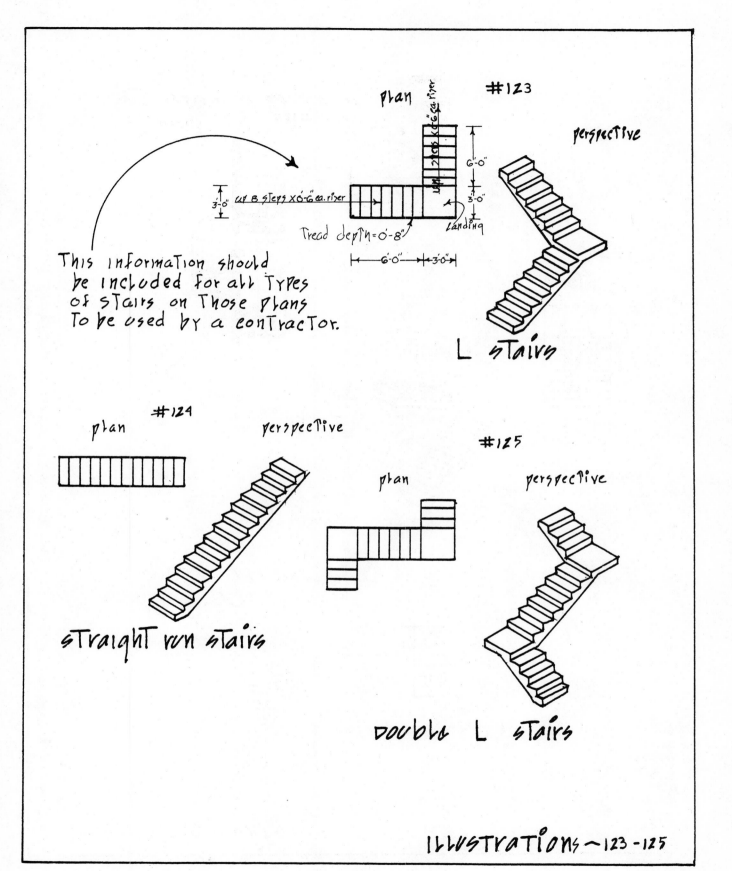

Plan

#123

perspective

up 2 steps X 0'-6" ea. riser

6'-0"

3'-0"

Landing

3'-0"

up 8 steps X 0'-6" ea. riser

Tread depth = 0'-8"

6'-0"

3'-0"

L stairs

This information should
be included for all Types
of Stairs on Those plans
To be used by a contractor.

#124

plan

perspective

straight run stairs

#125

plan

perspective

Double L stairs

ILLUSTrations ~ 123 - 125

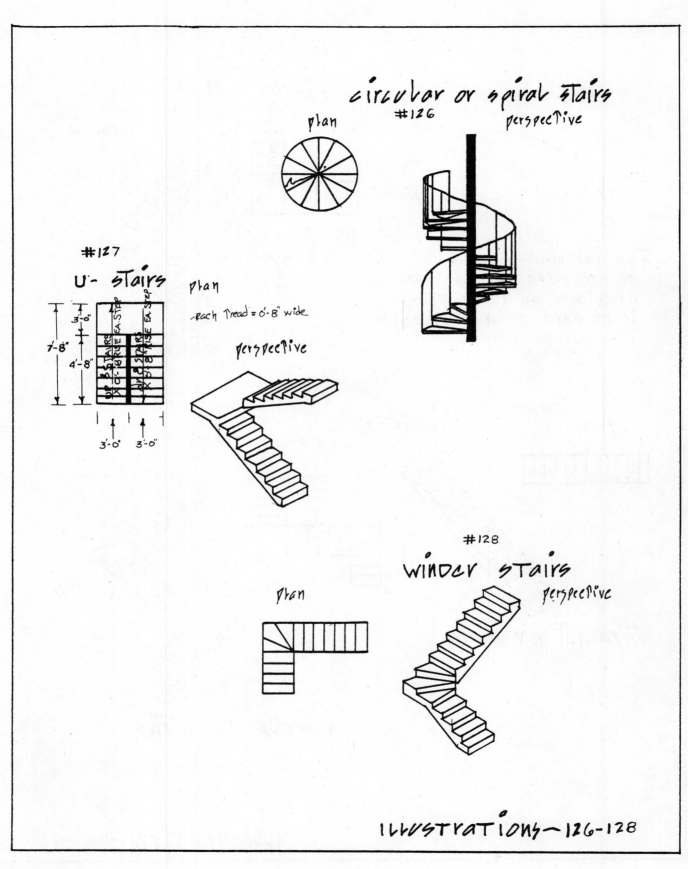

circular or spiral stairs
#126

plan

perspective

#127

U - stairs

plan

-each Tread = 6'-8" wide

perspective

3'-0"

7'-8"

4'-8"

3'-0"   3'-0"

#128

winder stairs

plan

perspective

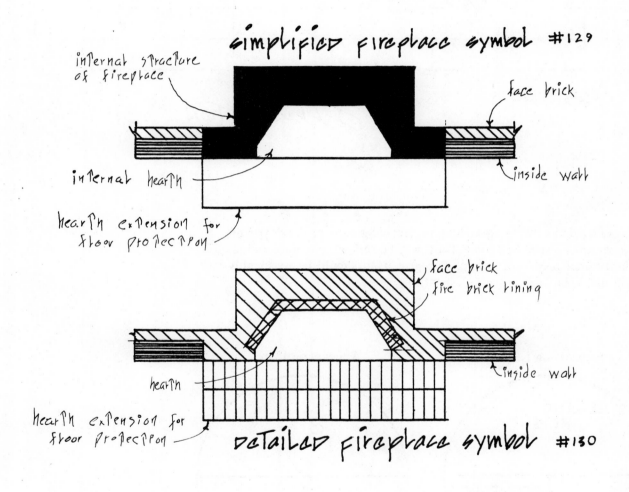

simplified fireplace symbol #129

internal structure of fireplace

face brick

internal hearth

inside wall

hearth extension for floor protection

face brick
fire brick lining

inside wall

hearth

hearth extension for floor protection

detailed fireplace symbol #130

ILLUSTRATIONS~129, 130

DOOR SCHEDULE					
Symbol (Key)	Quan.	Type	Door Size	Manufact. Number	Remarks
A	2	FLUSH	3'-0"X6'-8"	EF 368	1¾" SOLID BIRCH
B	1	BI-FOLD	6'-0"X6'-8"	BF 36AL	2 UNITS EA 36"W.

ILLUSTRATION #131

GRIDS

Grids are used mostly in commercial projects of great size. Grids permit easy location of a specified point. They may be compared to longitudinal and latitudinal points on a map. The grid itself should be reasonably light in line value. (See illustration #135.)

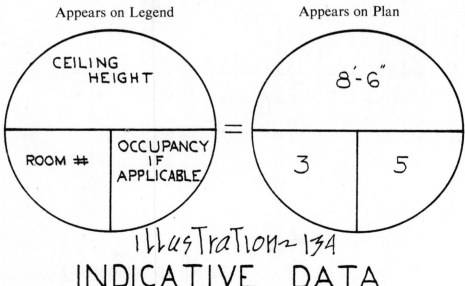

ILLUSTRATION 134

INDICATIVE DATA

FINISH SCHEDULE

IllustraTion ~ 132

RM. No.	LOCA-TION	WALLS					FLOOR					CEILING (Same Headings as Walls)
		Trt mnt	Man. Name	Man. No.	Color	Rmks	Trt mnt	Man. Name	Man. No.	Color	Rmks	
1	North	PAINT	B. MOORE	G 114	GRAY	2 COATS	CARPET	SAXONY	GS 12	GRAY SUEDE		
	South											
	East	MIRROR	P.P.G.	PE 27	GRAY TINT	4 PANELS 4'X10'						
	West	PAINT	B. MOORE	G 114	GRAY	2 COATS						
2	North	PAINT	B. MOORE	G 114	GRAY	2 COATS	SAME TREATMENT AS ROOM 1.					
	South											
	East	FABRIC	STROHEIM & ROMANN	GF-54W	GRAY FLANNEL							
	West	MIRROR	P.P.G.	PE 27	GRAY TINT	3 PANES 4'X10'						
	North											
	South											
	East											
	West											

CONTINUE AS NECESSARY

LEGEND

FINISH REFERENCE:

CARPET ① GRAY COLOR WILL BE SUPPLIED BY DESIGNER

VINYL ① WG VINYLS BUCKENEER SUEDE #BRS 516 C MAUVE

VINYL ② WG VINYLS VALIANT STRIPE #VT358IC PEARL GRAY

PAINT ① GRAY COLOR, TO MATCH VINYL 2 COLOR

PAINT ② MAUVE COLOR, TO MATCH VINYL ① COLOR

PAINT ③ STANDARD WHITE COLOR

MAXELL SPORTSWEAR
1384 B'WAY N.Y. N.Y.
3-1-81 SCALE: 1/4"=1'-0"
FINISH SCHEDULE #3 OF 5
W.E.M. DESIGNS DWN. BY: J.E.

FINISH SCHEDULE

SPACE NAME	NO	FLOOR	BASE-BOARD	WALL NORTH	SOUTH	EAST	WEST	CEILING	HT.
RECP'T	1	CARPET	NONE	VINYL ①	VINYL ①	VINYL ①	VINYL ①	ACOUSTICAL TILE	8'-6"
OFFICE	2			PAINT ①	PAINT ①	PAINT ①	PAINT ①		
OFFICE	3								
SHOWROOM	4			PAINT ②	PAINT ②	PAINT ②	PAINT ①		
SHOWROOM	5								
OFFICE	6						PAINT ③		
CORRIDOR	7			VINYL ①	VINYL ①	VINYL ①	VINYL ①		
SHOWROOM	8			PAINT ①	PAINT ①	PAINT ①	PAINT ①		
STORAGE	9			PAINT ②	PAINT ②	PAINT ②	PAINT ②		

illustration ~ 133

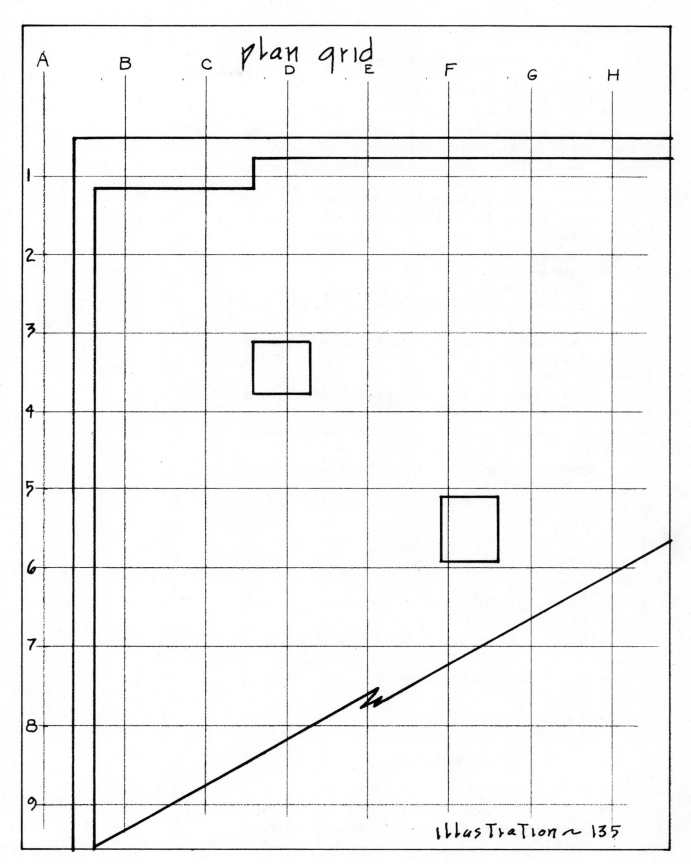

plan grid

ILLUSTRATION ~ 135

REVIEW OF IMPORTANT POINTS

1. Vary line weights for proper presentation.

2. Keep individual line weights uniform in appearance.

3. All lines have specific beginning and end points; be sure these points are obvious and not carelessly drafted.

4. When plans are to be reproduced in the various forms, such as blue, black, or brown line prints, photostatic copies, or negatives, be sure that necessary line weights are sufficiently heavy so that they can be reproduced well and that drafting paper is as transparent as possible.

5. Be precise and consistent in:
   a. Measurement indication;
   b. Architectural lettering;
   c. Graphic representation of doors, walls, and all architectural elements.

6. Use short or long dash lines to indicate above-eye-level architecture, e.g., skylight or beams. Use both when necessary to avoid confusion due to proximity of elements.

7. Use dotted and dash lines to indicate a partially hidden object, e.g., chair seat under a table.

8. Give additional explanation via leader arrows as may be required.

9. Use proper pouchering techniques to indicate composition of architectural elements.

10. Indicate direction, tread depth, and riser heights of all platforms and steps.

11. Supply as much information on schedules as may be necessary.

12. Use breaklines ($\frac{1}{7}$) when necessary.

13. Wash hands frequently to help avoid smudging.

14. Work from the top of the paper downward to help eliminate smudging.

15. Have adequate lighting that does not create glare on drafting surface.

16. Keeping in mind the type and purpose of the plans, emphasize that aspect when drafting.

18. Indicate means of main entry.

QUIZ QUESTIONS

1. List the three categories of architectural lettering.

2. What are the normal heights of lettering and when is each used?

3. What are the three types of line weights? Give their definitions.

4. Show the seven means of measurement indication.

5. What are the two units in which measurements must be indicated?

6. On what axis should measurement size be noted?

7. When are break lines used?

8. When are leader arrows used?

9. What is pouchering and when is it used?

10. Illustrate sixteen methods of pouchering and name their indications.

11. Name ten types of doors and draft a plan and frontal view of each.

12. How are above-eye-level elements indicated?

13. How are hidden elements indicated?

14. Name ten types of windows and draft their plan and frontal views.

15. Name six types of stair cases and draft their plan and perspective view.

16. What is the difference between a simplified fireplace symbol and a detailed fireplace symbol? When is each used?

17. List the type of information found on a door schedule.

18. List the type of information found on a window schedule.

19. List the type of information found on a finish schedule.

20. What is meant by indicative data?

EXERCISES

1. Using HB lead, draft twenty-five horizontal lines in guideline weight, each line 15'-0" in ¼" scale.

2. Using F lead, draft twenty-five horizontal lines in guideline weight, each line 15'-0" in ¼" scale.

3. Using H lead, draft twenty-five horizontal lines in guideline weight, each line 15'-0" in ¼" scale.

4. Using 2H lead, draft twenty-five horizontal lines in guideline weight, each line 15'-0" in ¼" scale.

5. Using 4H lead, draft twenty-five horizontal lines in guideline weight, each line 15'-0" in ¼" scale.

6. Using HB lead, draft twenty-five horizontal lines in medium weight, each line 15'-0" in ¼" scale.

7. Using F lead, draft twenty-five horizontal lines in medium weight, each line 15'-0" in ¼" scale.

8. Using H lead, draft twenty-five horizontal lines in medium weight, each line 15'-0" in ¼" scale.

9. Using 2H lead, draft twenty-five horizontal lines in medium weight, each line 15'-0" in ¼" scale.

10. Using 4H lead, draft twenty-five horizontal lines in medium weight, each line 15'-0" in ¼" scale.

11. Using HB lead, draft twenty-five horizontal lines in heavy weight, each line 15'-0" in ¼" scale.

12. Using F lead, draft twenty-five horizontal lines in heavy weight, each line 15'-0" in ¼" scale.

13. Using H lead, draft twenty-five horizontal lines in heavy weight, each line 15'-0" in ¼" scale.

14. Using 2H lead, draft twenty-five horizontal lines in heavy weight, each line 15'-0" in ¼" scale.

15. Using 4H lead, draft twenty-five horizontal lines in heavy weight, each line 15'-0" in ¼" scale.

16. Using HB lead, draft twenty-five vertical lines in guideline weight, each line 15'-0" in ¼" scale.

17. Using F lead, draft twenty-five vertical lines in guideline weight, each line 15'-0" in ¼" scale.

18. Using H lead, draft twenty-five vertical lines in guideline weight, each line 15'-0" in ¼" scale.

19. Using 2H lead, draft twenty-five vertical lines in guideline weight, each line 15'-0" in ¼" scale.

20. Using 4H lead, draft twenty-five vertical lines in guideline weight, each line 15'-0" in ¼" scale.

21. Using HB lead, draft twenty-five vertical lines in medium weight, each line 15'-0" in ¼" scale.

22. Using F lead, draft twenty-five vertical lines in medium weight, each line 15'-0" in ¼" scale.

23. Using H lead, draft twenty-five vertical lines in medium weight, each line 15'-0" in ¼" scale.

24. Using 2H lead, draft twenty-five vertical lines in medium weight, each line 15'-0" in ¼" scale.

25. Using 4H lead, draft twenty-five vertical lines in medium weight, each line 15'-0" in ¼" scale.

26. Using HB lead, draft twenty-five vertical lines in heavy weight, each line 15'-0" in ¼" scale.

27. Using F lead, draft twenty-five vertical lines in heavy weight, each line 15'-0" in ¼" scale.

28. Using H lead, draft twenty-five vertical lines in heavy weight, each line 15'-0" in ¼" scale.

29. Using 2H lead, draft twenty-five vertical lines in heavy weight, each line 15'-0" in ¼" scale.

30. Using 4H lead, draft twenty-five vertical lines in heavy weight, each line 15'-0" in ¼" scale.

31. On fifteen 8½" x 11" sheets of blank paper, draft horizontal guidelines, alternating the space between these lines at ⅛" and

¾₆″ intervals. Print an article from the newspaper on five of these sheets using the ¾₆″ spaces as the writing surface and the ⅛″ spaces as spaces; do this in back slant printing. Repeat this exercise on five additional sheets in vertical printing, and finally, do this exercise on the remaining five sheets in forward slant printing.

32. Draft the following dimensions, using each of the measurement indications illustrated in the text: in ¼″ scale—5′–2″, 4′–9″, 10′–6″, 11′–4″, 2′–1″, 20′–7″, 8′–4″, 19′–3″, and 44′–8″; in ⅛″ scale—5′–2″, 4′–9″, 10′–6″, 11′–4″, 2′–1″, 20′–7″, 8′–4″, 19′–3″, and 44′–8″; in 3″ scale—1′–2″, 5′–4″, 2′–1″, 7′–9″, 4′–3″, 6′–7″, 3′–11″, 10′–5″, and 8′–2″; in 1″ scale—1′–2″, 2′–1″, 5′–6″, 6′–5″, 4′–4″, 3′–7″, 9′–8″, 8′–9″, and 10′–11″.

33. Draft ten types of doors illustrated in the text. Draft at the scale in which they are illustrated.

34. Draft ten types of windows illustrated in the text. Draft at the scale in which they are illustrated.

35. Draft six types of stair cases, as shown in the text. Be sure to indicate all necessary information, such as riser height, tread depth, direction, and width of stair case.

3
THE VARIOUS TYPES OF PLANS

PREPARATION OF DRAFTING PAPER

There are no absolute rules for a format under which drawings must be made. Hence, it is difficult to establish criteria by which a drawing's visual acceptability, and therefore degree of usefulness, can be measured. We do know, however, that a good drawing can lose a great portion of its effectiveness when not presented in its best light and format, and in its highest degree of clarity. Therefore, the following suggestions for proportionate measurement are offered in order to create a framework around the drawn plan itself and the elements that it comprises. The visual impact of the drawing will in turn be increased, and therefore amplification of its comprehension will be accomplished. By using these recommended spacings, not only will attention be focused on the drawing itself, but a clear and attractive graphic presentation will result. (See illustration #136.)

There are, in addition, drafting sheets that can be bought in an art supply store that already have borders, a legend box, and a title box pre-printed on them. One may even have customized sheets printed with one's company name or any other additional information desired. These, of course, will save enormous amounts of time and effort for the draftsperson.

These suggested measurements, when applied to varied size sheets of drafting paper, will require adjustment so that their proportional value remains relative to the size sheet used.

THE LEGEND BOX

The legend box is used to further explain any graphic notations or symbols that have been used in the body of the drawing. The legend box is placed to the right of the drawing. One can refer to the legend box when there are graphic notations or symbols used in the drawing that require further explanation, For instance, in the case of a construction plan, the various types of pouchering (the method of indicating the particular material of which an architectural element is composed, e.g., concrete, glass, plywood, brick, stone, etc.) would not only appear on the drawing, but in addition, the symbols and their description, definition, or explanation would appear in the legend box. In the case of an electrical plan, the symbol that has been used to represent a certain type of outlet or switch in the drawing would also appear in the legend with its description, definition, or explanation. The same principles are true for

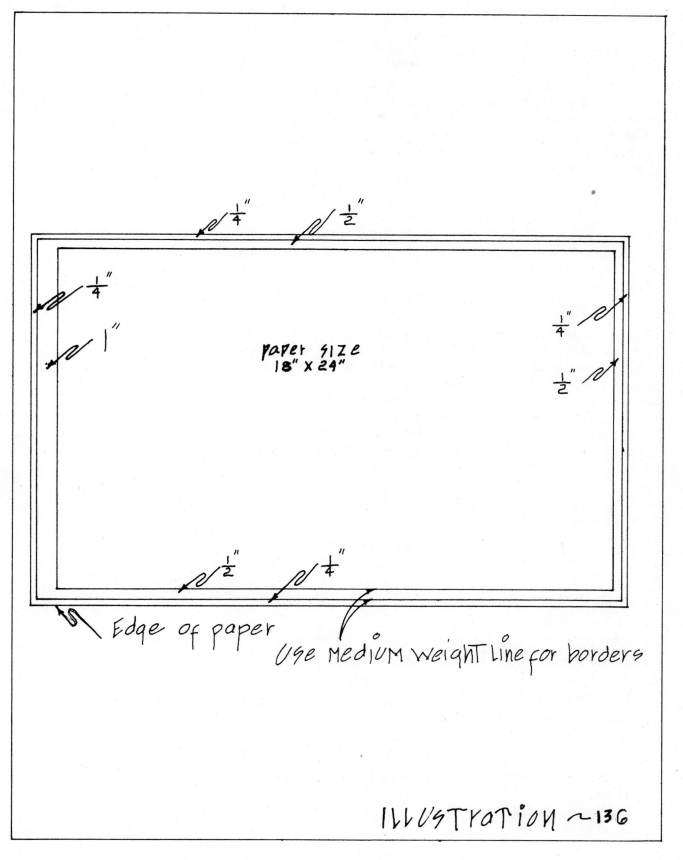

paper size
18" X 24"

Edge of paper

Use medium weight line for borders

ILLUSTRATION ~136

reflected ceiling or any other type of plan. (See illustrations #137 and #138.)

Blanket statements (statements that are indicative of general information) would also be placed in the legend box. They should, however, be written in such a way that they are not capable of being confused with descriptions, explanations, or definitions of graphic symbols. This may be accomplished by indenting the blanket statement. Some possible blanket statements that may be seen are:

"All measurements to be verified by tradesmen in the field";
"All windows located 3'–0" above finished floor";
"All measurements indicated are to the center of the element";
"All interior walls are 0'–5" thick";
"Designer to be notified prior to any changes being made in the field"; and
"All construction to meet building code requirements."

In essence, any statements that could be more easily read or more clearly understood in the legend box than on the drawing paper should be placed in this location. One may also use a blanket statement in the legend box instead of the drawing proper so that additional attention is drawn to the statement. Finally, an additional purpose of placing a blanket statement only in the legend box is to negate the necessity of the statement being placed repeatedly in several locations of the drawing proper. Placing it once in the legend box avoids excessive writing on the drawing and therefore facilitates its readability. (See illustration #139.)

THE TITLE BOX

A title box is used to identify general information on the drafted sheet such as the following:

1. Name of client;
2. Date of drawing;
3. Type of drawing (e.g., furniture plan, electrical plan, etc.);
4. Drawing number (if there are several drawings, i.e., 1 of 6, 2 of 6, 3 of 6, etc.) in order to indicate a complete set intact;
5. Name of draftsperson;
6. Name of company;
7. Scale used;
8. Revision date (if applicable), which insures that all parties concerned are discussing the same drawing if a drawing has been changed; and
9. Signature approval of person initiating plans. (See illustration #140.)

Legend box within borders

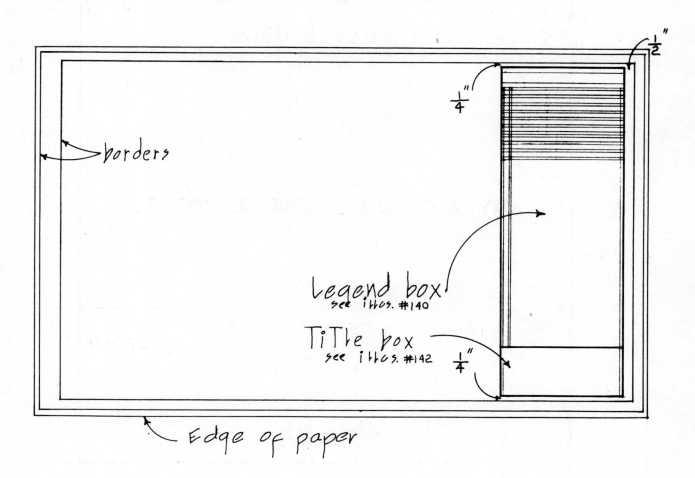

borders

Legend box
see illus. #140

Title box
see illus. #142

$\frac{1}{4}''$

$\frac{1}{4}''$

$\frac{1}{2}''$

Edge of paper

ILLUSTRATION ~137

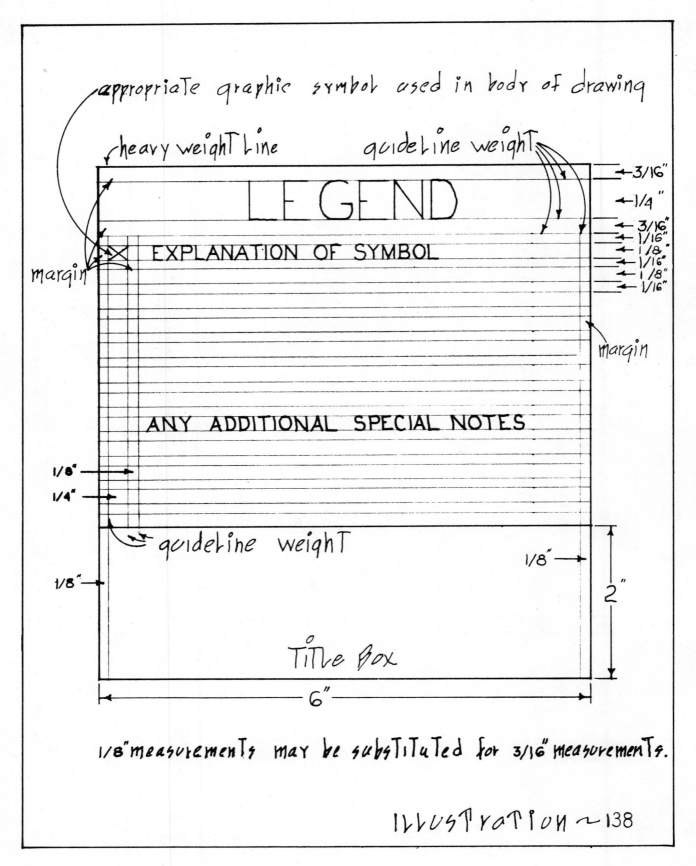

appropriate graphic symbol used in body of drawing

heavy weight line   guideline weight

← 3/16"

← 1/4"

← 3/16"
← 1/16"
← 1/8"
← 1/16"
← 1/8"
← 1/16"

LEGEND

EXPLANATION OF SYMBOL

margin

margin

ANY ADDITIONAL SPECIAL NOTES

1/8" →

1/4" →

guideline weight

1/8" →   1/8" →

1/8" →

2"

TITLE BOX

6"

1/8" measurements may be substituted for 3/16" measurements.

ILLUSTRATION ~ 138

LEGEND

░ FACE BRICK
⠿ TERRAZO
⤬ MARBLE
▨ CLAY TILE

NOTES:
   ALL INSIDE WALLS = 0'-5" THICK
   ALL OUTSIDE WALLS = 0'-6" THICK
   CEILING HEIGHT = 8'-0"
   WINDOW LEVEL = 2'-0" A.B.F.
   WINDOW HEIGHT = 6'-0" A.B.F.

EXISTING CONDITIONS PLAN
C.H. FAIRCHILD / 3-1-81 / SCALE: 1/4"=1'-0"
PREPARED BY J.M. FLATT
FOR W.E.M. DESIGNS / PAGE 1 OF 10

Illustration 139

TITLE BOX

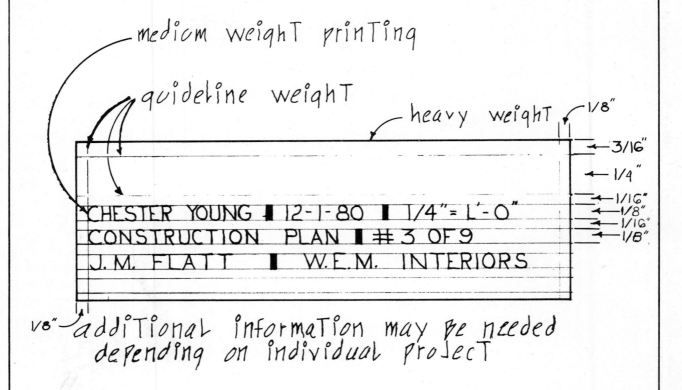

medium weight printing

guideline weight

heavy weight

1/8"

3/16"

1/4"

1/16"
1/8"
1/16"
1/8"

CHESTER YOUNG ▌ 12-1-80 ▌ 1/4"= 1'-0"
CONSTRUCTION PLAN ▌ #3 OF 9
J. M. FLATT ▌ W.E.M. INTERIORS

1/8" additional information may be needed
depending on individual project

1/8" measurements may be substituted for 3/16" measurements.

ILLUSTRATION ~140

The arrangement of the information given is a matter of personal style or preference. Usually, the most important information, such as type of plan or company name, is given first. However, this is not to be considered a cardinal rule.

UNDERSTANDING PLANS

Most of us have become accustomed to viewing an object by looking directly at it (frontal view). However, when drafting plans, such as existing conditions, demolition, construction, electrical, reflected ceiling, and furniture plans, we are doing so from a "bird's-eye view." That is to say, we are viewing objects and rooms from a reference point directly above it and looking straight downward (as though we are suspended just below ceiling level, so that ceiling fixtures are not considered in this view.) The exception, however, would be a reflected ceiling plan, combination reflected ceiling/electrical plan, or a complicated existing conditions plan. These plans are viewed as though the ceiling were transparent and one were looking downward from above ceiling level. Keeping this principle in mind, one should not have any trouble in being able to draft or understand a plan. Because of the view from which plans are seen ("bird's-eye"), not all information graphically displayed can be made completely clear to the reader of the plan, for example, beams, archways, and sometimes window and door types. There, one must employ leader arrows with additional explanation, or for doors and windows one must use a schedule, as has been discussed previously. In addition, one may have to use elevations, sections, and detail drawings, which will be discussed in a future chapter.

Usually, each type of plan is drafted on a separate sheet of paper. However, when plans are relatively simple, it is considered acceptable and sometimes preferable to combine related plans. For example, a demolition plan and construction plan, when drafted as one plan, would become a demolition/construction plan (to be discussed later in this chapter). An electrical plan and a reflected ceiling plan, when drafted as one plan, would become an electrical/reflected ceiling plan. (When this is done, all information pertinent to each plan must be included.) On more complicated projects and when more advanced drafting is understood, an existing conditions plan will not only include existing architecture, but will include existing electrical information or other required elements.

Plans may be categorized in an over-simplified manner by dividing them into three broad categories: (1) Those plans given to a contractor, such as existing conditions, demolition, construction, reflected ceiling, electrical, and architectural elevation, section, or detail plans; (2) Those plans given to a cabinetmaker, such as a cabinetry elevation, section, or detail; and (3) Those plans given to a client, such as dressed or undressed furniture, interior elevation of a design concept, or detail drawing of a custom piece of furniture plan. The various types of plans that are explained below are listed in their usual order of occurrence.

THE EXISTING CONDITIONS PLAN

The existing conditions plan is that plan that indicates those architectural conditions or elements that are present in a given project prior to any demolition or construction work being performed. In addition to indicating the composition of any structural elements, it indicates *all* measurements and makes note of any unusual architectural features, for example, beams, doors, fenestration (windows), columns, archways, etc. The purpose of an existing conditions plan is to familiarize the architect, designer, and contractor with the architectural characteristics of the project. It is the plan to which a contractor will refer back if there are any questions concerning the subsequent plans. Therefore, it is imperative to include *all information*—all architecture and all measurements. In doing so, each person can perform his or her duties and responsibilities more efficiently. The existing conditions plan also serves as the basis for drafting all subsequent plans. On an existing conditions plan, all architectural elements that currently exist are considered of primary importance and should therefore be drafted in heavy line weight. All other notations, measurement indications, etc. would be drafted in medium line weight. There should be a clearly discernible difference between the two line weights. One should not be afraid to print as much explanation of a particular element as may be required to insure what one has drafted or wishes to convey.

(See illustration #141 for a simple existing conditions plan and illustration #142 for a more complicated existing conditions plan that includes other elements, such as those included in reflected ceiling and electrical plans. Note the difference between the two plans, and keep in mind that separate plans are preferable whenever possible for purposes of clarity. The more complicated existing conditions plan can at times be very confusing and therefore should be avoided if possible.)

THE DEMOLITION PLAN

The demolition plan is drafted as an overlay of the existing conditions plan. An overlay is a sheet of drafting paper that when placed on top of the original drawing, permits the original drawing to be seen through it. It literally lays over the original drawing. This permits one to trace those areas necessary to execute the drafting of the subsequent plan without having to re-measure established elements.

It gives all measurements and notations necessary in order to note what structure is to be demolished, in addition to any reference points necessary to begin the demolition of a specified area (e.g., how far along a wall demolition is to take place). However, instead of indicating the composition of structural elements, it calls the attention of the contractor to those structural elements of the project that must be demolished. This is accomplished by the draftsperson devising a pouchering symbol to represent demolition and placing it in the outline of the structural elements to be demolished. (The demolition symbol replaces the symbol

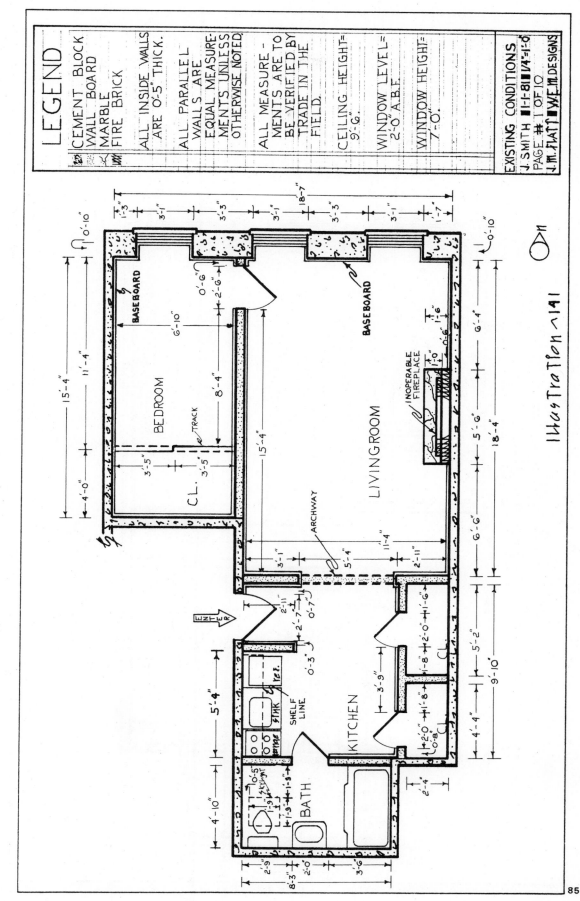

Illustration 141

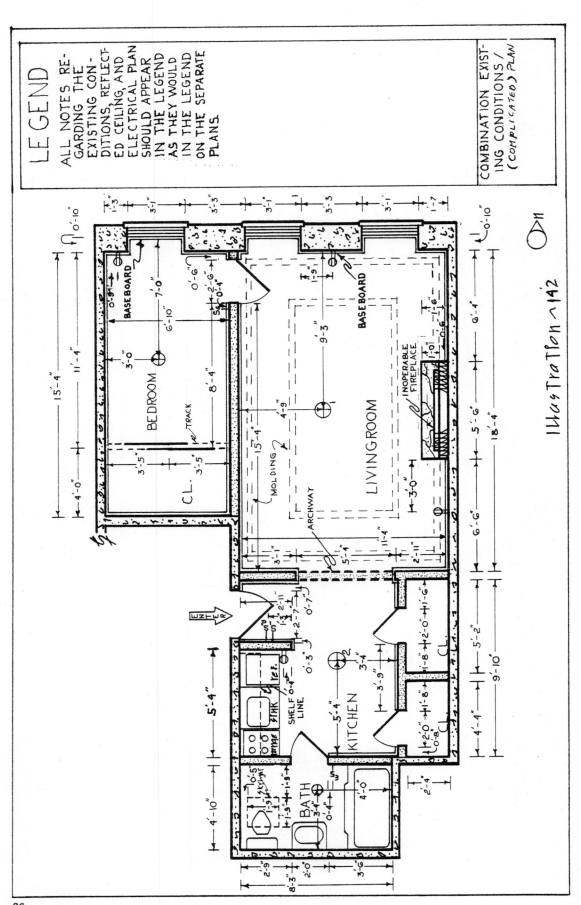

that had been used to indicate the material composition of an element on the existing conditions plan.) The type of demolition symbol used is arbitrary. It should, however, be a symbol that has not been, nor will be, used on any other plans and one that can be clearly read. The symbol must appear on the body of the drawing and, of course, in the legend box along with a definition. Any special notations necessary may also be put in the legend box. Since those areas to be demolished are of primary importance, one would use heavy line weight in specifying these areas. (See illustration #143.)

Because this plan, in addition to most other plans, may be used by a contractor to bid for a job (give an estimate of its cost), it is important that one be most clear and explicit in drafting those elements to be demolished and give all necessary peculiarities of the demolition, for example, "demolition to begin 3′–0″ above finished floor" or "archway in wall to be demolished has 7′–0″ clearance and is 5′–4″ wide," or to indicate that composition of element to be demolished is of a special material, such as concrete or metal. In this way, a contractor may take into consideration all aspects of the required work and can therefore bid fairly.

If elements that are being demolished are to be reused during construction, it would be necessary to indicate such a request in the legend box as a blanket statement, for example, "kitchen cabinets to be rehung on west wall," or "existing doors to be rehung in new frames," and so forth. Again, one should not be hesitant about giving as much written explanation as may be required.

A total measurement of each area may be given unless a breakdown of measurements is required for bidding purposes (cost estimates), such as in the case of a wall to be demolished that includes an archway. (See illustration #143.)

THE CONSTRUCTION PLAN

The construction plan, in essence, is the opposite of the demolition plan. The point of a construction plan is to instruct the contractor where he is to build walls, doors, windows, platforms, steps, fireplaces, and any other structural elements. In addition, it specifies the length, height, material composition, and very often the method of construction of these elements. Again, this may be done in the overlay manner, substituting construction symbols in place of demolition symbols. Do not, however, use the same symbol for demolition and construction. In order for a contractor to do his job properly, one must be very explicit and exacting when giving all necessary measurements and notations of these elements. Be sure to also include any necessary reference points. Nothing may be assumed. Since construction elements are of primary importance, these elements must be drafted in heavy line weight. It is especially imperative that sufficient explanations have been supplied on this drawing for the contractor's clarification. Misinterpretation will lead to mistakes, and mistakes can be very costly both from a time and from a financial point of view. (See illustration #144.)

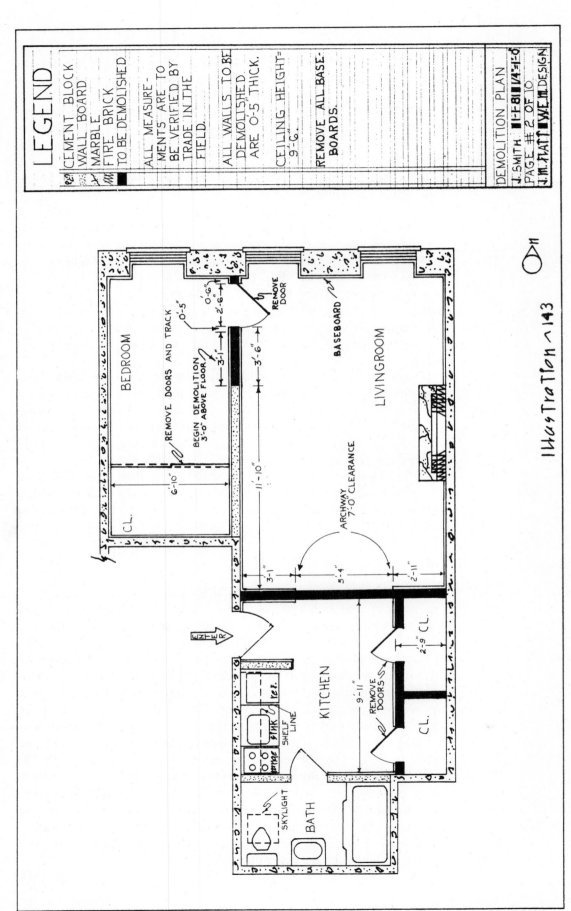

Illustration ~143

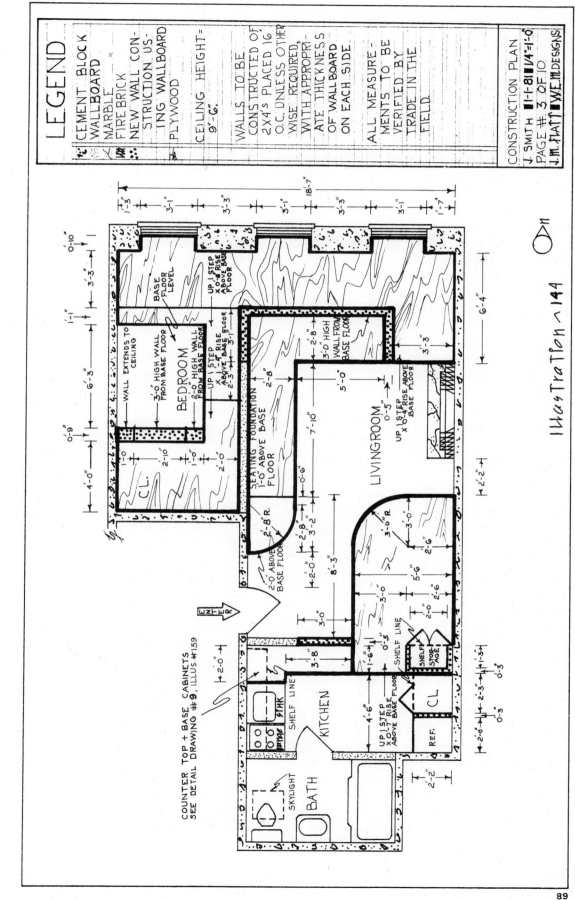

Illustration ~ 144

Sometimes, a single demolition/construction plan may be devised instead of separate demolition and construction plans. This single demolition/construction plan may be used only when very simple contract work is to be performed. It may not be used when one element is to be constructed in the same space that demolition must occur, such as when a door is to be installed in the same space where a window currently exists.

In the case of a demolition/construction plan, both demolition and construction are considered of primary importance. Each would require heavy line weight.

Note: Prior to any demolition or construction, a designer should consult an architect to make certain that the design changes are feasible from the structural and coding law points of view. An architect has the necessary training and awareness of building codes necessary to determine structural feasibility and coding legality of residential and commercial demolition and construction changes. If there are any questions in the designer's mind, he or she would be well advised to consult an architect. In doing so, he or she would avoid many problems that could prove most disastrous.

Remember, however to indicate what symbols are to be used for construction and which ones are being used for demolition. (See illustration #145.)

THE FURNITURE PLAN

Most of the plans that are drafted are for a contractor's use or cabinet-maker's use. The furniture plan is but one exception. This plan is devised for the client. Because of its differing use, the furniture plan is somewhat less technical. If measurements need be given (they usually need not), only the measurement of furniture (in inches) and the overall dimensions of each room in feet and inches would normally appear. It would also be wise to label each room.

Furniture need not be labeled unless an item is a custom piece and not easily recognizable as such on the plan. In that case, one has the option of labeling the item directly on the body of the drawing (which is preferable) or keying it to the legend. (See illustration #146.)

A furniture plan may be relatively simple or may be "dressed" (see Chapter 5, *Dressing a Plan*). Whichever is more suitable for the occasion would be dictated by the size of the project, needs of the client, and preferences of the designer.

A furniture plan would include all furniture in addition to floor and table lamps, and architectural elements such as stairs, fireplaces, platforms, and so on. The furniture of a room, being the most important elements of the plan, should be outlined in heavy line weight.

The pouchering method one uses for the architectural elements in a furniture plan is usually the non-specified method, as indicated in illustration #101. In addition, if it is important to indicate a special aspect of the design concept to the client, such as glass wall, different height

DEMOLITION / CONSTRUCTION

LEGEND

■ TO BE DEMOLISHED

▨ TO BE CONSTRUCTED

▨ 8" BLOCK

▒ WALLBOARD

ALL MEASUREMENTS TO BE VERIFIED BY TRADE IN THE FIELD.

WALLS TO BE CONSTRUCTED OF 2X4's PLACED 16" O.C. UNLESS OTHERWISE REQUIRED, WITH APPROPRIATE THICKNESS OF WALLBOARD ON EACH SIDE.

F. BAKAL RESIDENCE   SCALE: 1/4"= 1'-0"
1-1-81
DEMOLITION / CONSTRUCTION
J.M. FLATT

LIVINGROOM

CL

CL

ARCHWAY

91

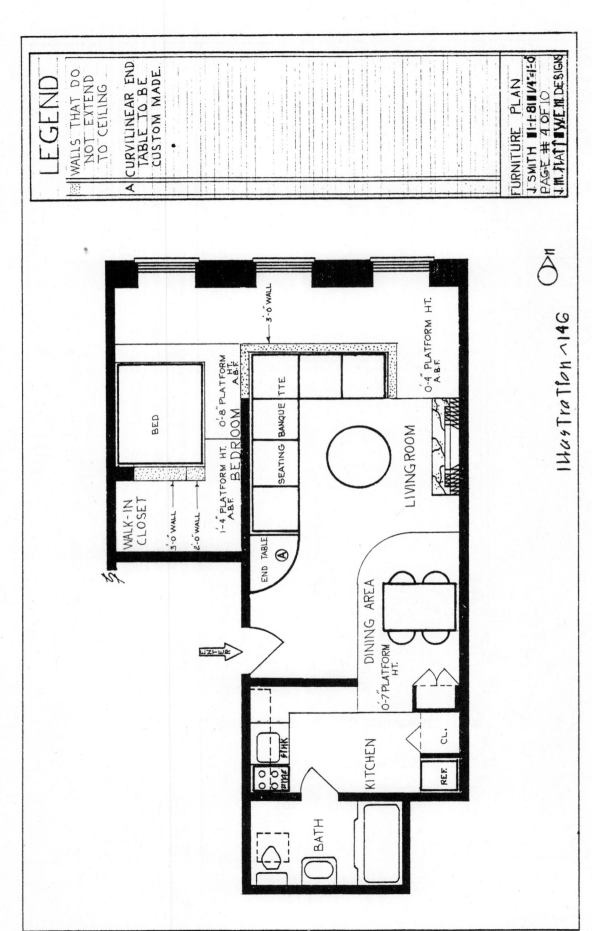

LEGEND

WALLS THAT DO
NOT EXTEND
TO CEILING

A CURVILINEAR END
TABLE TO BE
CUSTOM MADE.

FURNITURE PLAN
J. SMITH 1-1-81 1/4"=1'-0"
PAGE # 4 OF 10
J.M. PLATT W.E.M. DESIGNS

WALK-IN
CLOSET

3'-0 WALL
2'-0 WALL

BED

BEDROOM

1'-4 PLATFORM HT.
A.B.F.

0'-8 PLATFORM
HT.
A.B.F.

3'-0 WALL

SEATING BANQUETTE

END TABLE (A)

LIVING ROOM

0'-4 PLATFORM HT.
A.B.F.

DINING AREA

0'-7 PLATFORM
HT.

ENTER

BATH

SINK

KITCHEN

CL.

REF.

Illustration ~146

walls, or floor covering, it is suggested that one employ a leader and notation.

THE REFLECTED CEILING PLAN

The reflected ceiling plan is what it suggests in its title—a reflection of the ceiling. This concept is easier to understand if one were to imagine the ceiling being transparent and one were looking downward through this transparency, therefore being able to see all ceiling elements, such as beams and lighting fixtures. This plan instructs the contractor where to place luminaires (lighting fixtures) that are to be recessed into, or surface-mounted on, the ceiling. In addition, it indicates any overhead beams, HVAC vents (Heating, Ventilation, Air Conditioning vents), and ceiling tiles or molding, if applicable. In showing the wall architecture of a reflected ceiling plan, one does not show doors or windows unless they continue to the ceiling. Instead, indication is made to show these elements as solid wall since one is seeing through the ceiling to the wall points and not to the door or window points.

The two most important requirements of this plan are (1) to specify, in the legend box, what symbol has been used to represent a particular luminaire, giving manufacturer's name, number, finish, dimensions, lamp (light bulb) size, type, wattage, lens and louvre requirement, and (2) to indicate on the body of the drawing the location of the luminaire by giving two perpendicular measurements of distance from stationary architectural points.

However, if for instance, you want to locate three luminaires equidistant between two parallel walls, one need only give the distance away from the third perpendicular wall and use the symbol "E.Q." (equidistant) between the parallel walls and three fixtures. (See illustration #147.) Usually, the measurement given is indicated to the center of the fixture and not to the edge.

The reflected ceiling plan may be done as an overlay of the furniture plan. However, only architectural or structural elements would be indicated and furniture would not. The reflected ceiling plan is drafted after the furniture plan to insure that the most appropriate lighting solutions have been devised. Heavy line weight would be used for all indications of ceiling elements. (See illustration # 148.)

THE ELECTRICAL PLAN

Usually, an electrical plan is used in conjunction with a reflected ceiling plan, even though each serves a distinct purpose. If each plan is relatively simple, one may combine both, and hence a reflected ceiling/electrical plan is created. However, if the plans are somewhat complicated, two separate plans must be drafted. The electrical plan indicates placement (location), type, and function of all outlets, switches, telephone jacks, and special electrical requirements in addition to the circuitry (which switches control which luminaires or outlets).

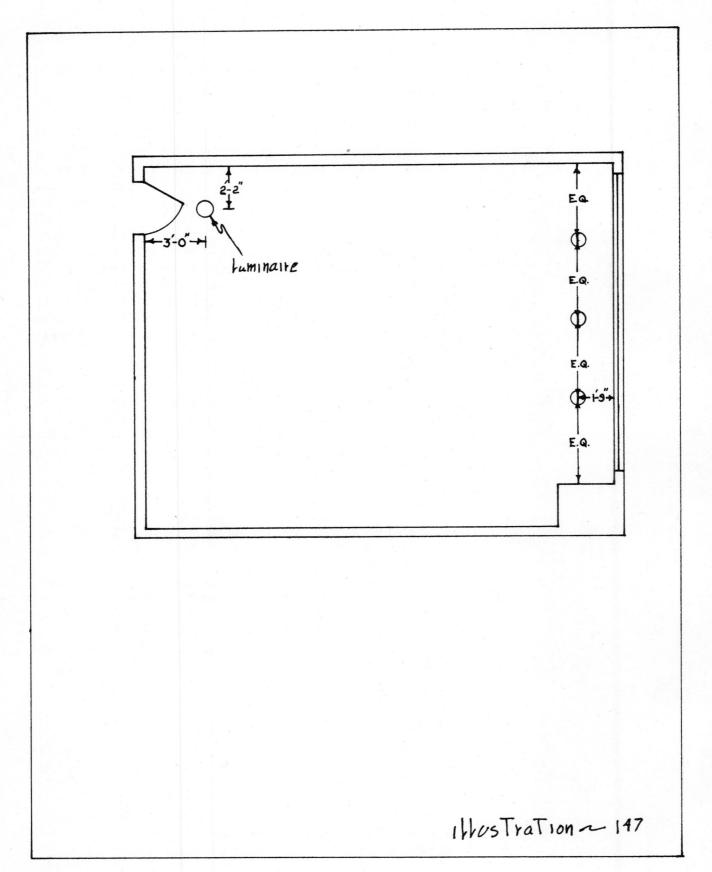

2-2"

E.Q.

3'-0"

luminaire

E.Q.

E.Q.

1'-9"

E.Q.

illustration — 147

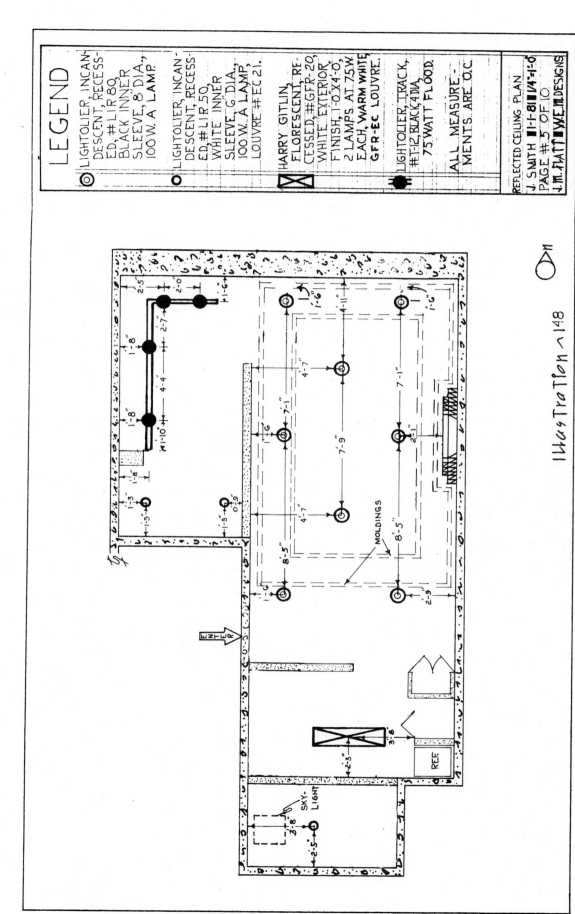

Illustration ~ 148

ELECTRICAL SYMBOLS

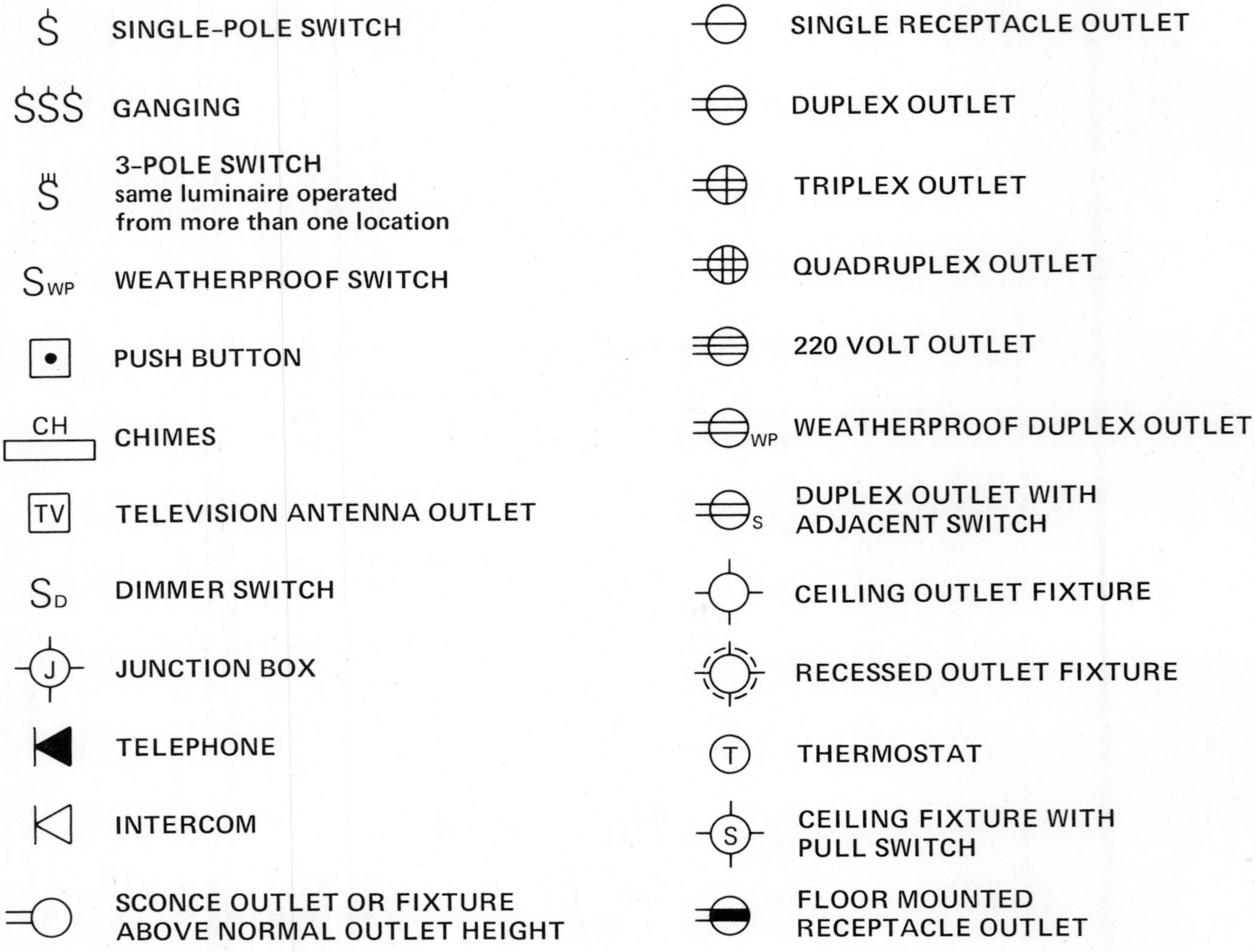
SINGLE-POLE SWITCH
GANGING
3-POLE SWITCH
same luminaire operated
from more than one location
S WP
WEATHERPROOF SWITCH
PUSH BUTTON
CH
CHIMES
TV
TELEVISION ANTENNA OUTLET
S D
DIMMER SWITCH
J
JUNCTION BOX
TELEPHONE
INTERCOM
SCONCE OUTLET OR FIXTURE
ABOVE NORMAL OUTLET HEIGHT
SINGLE RECEPTACLE OUTLET
DUPLEX OUTLET
TRIPLEX OUTLET
QUADRUPLEX OUTLET
220 VOLT OUTLET
WP
WEATHERPROOF DUPLEX OUTLET
S
DUPLEX OUTLET WITH
ADJACENT SWITCH
CEILING OUTLET FIXTURE
RECESSED OUTLET FIXTURE
T
THERMOSTAT
S
CEILING FIXTURE WITH
PULL SWITCH
FLOOR MOUNTED
RECEPTACLE OUTLET

illustration~149

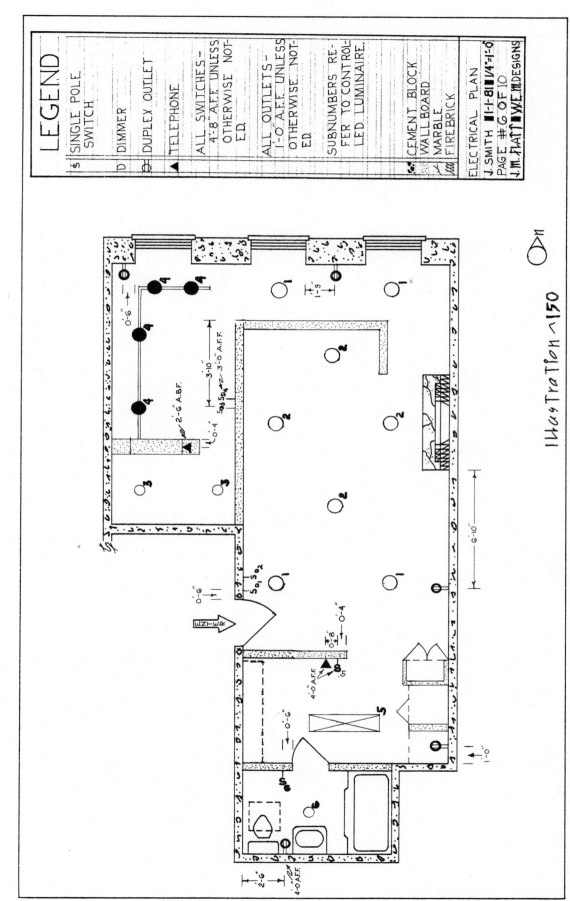

Illustration ~150

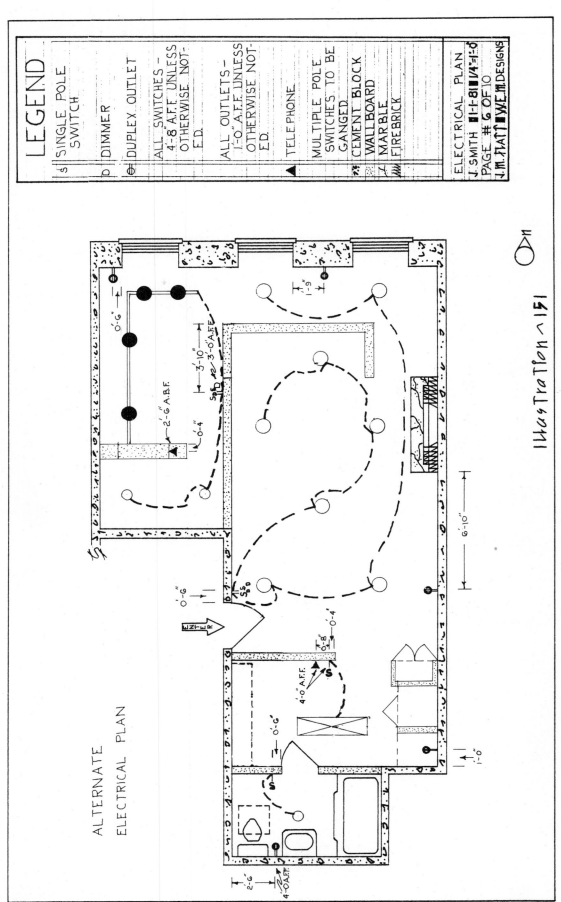

ALTERNATE
ELECTRICAL PLAN

Illustration 151

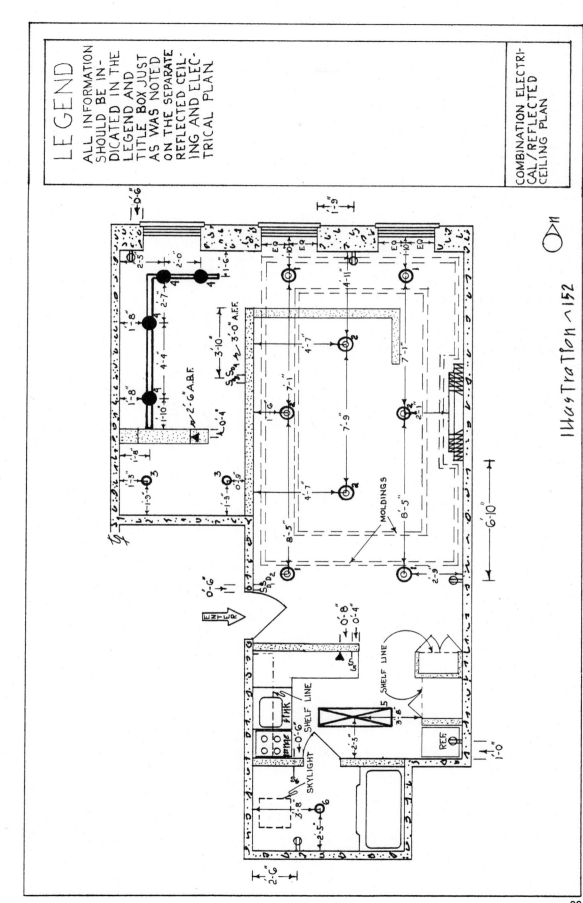

LEGEND

ALL INFORMATION SHOULD BE INDICATED IN THE LEGEND AND TITLE BOX JUST AS WAS NOTED ON THE SEPARATE REFLECTED CEILING AND ELECTRICAL PLAN

COMBINATION ELECTRICAL/REFLECTED CEILING PLAN

Illustration 152

There are several ways in which one may accomplish this. The following is considered the simplest method. Using the standard symbols given (see illustration #149), place the symbol in the body of the drawing at the desired location. Be sure to indicate any measurements that may be required so that the electrician can install the electrical element where specified. An electrician needs at least two measurements perpendicular to each other to install an electrical element. These would be how far A.F.F. (above finished floor) he is to install the outlets or switches (notes in the legend) and how far away from a stationary architectural element the electrical element is to be installed. One must then key each switch to the element (such as a luminaire) it is to control by placing a subscript number next to the switch and the controlled element. Again, explain in the legend box what each symbol represents. Switches, outlets, circuiting indications would of course be drafted in heavy line weight, as would be suggested by the titles of the plan. (See illustration #150.)

The alternative method (that method other than the subscript method that has just been explained) is to connect all switches to the luminaire or outlets that they control by use of a dotted line. (See illustration #151.)

When one combines a reflected ceiling and electrical plan, care must be taken to have all measurements located in such a way that they do not cause confusion as to which is a measurement for a switch and which is a measurement for a luminaire or ceiling tile. (See illustration #152.)

REVIEW OF IMPORTANT POINTS

1. Use margins around your drawing and in your legend and title boxes.

2. The legend box begins at the upper right hand corner of a drawing and illustrates any symbol that may have been used in the body of a drawing, in addition to giving a written explanation of the symbol. A legend box may include additional footnotes, if necessary, to a contractor.

3. A title box gives all necessary general information pertinent to a project. (See list under title box explanation, p. 78.)

4. Most plans are viewed from a bird's-eye view and may need additional notations with connecting leader arrows for further explanations.

5. Usually, each plan is drafted on a separate sheet of paper. However, some plans (e.g., demolition/construction and electrical/reflected ceiling) may be combined.

6. The three broad categories of plans are:
   a. For contractor (existing conditions, demolition, construction, reflected ceiling, electrical, architectural elevation, section, and detail plans);

    b. For client (undressed furniture, dressed furniture, interior design concept, elevation, and detail plans); and

    c. For cabinetmaker (cabinet section, elevation, and detail plans).

7. The following denote the most common types of plans listed in usual order of occurrence:
   a. Existing conditions;
   b. Demolition;
   c. Construction;
   d. Furniture (undressed or dressed);
   e. Reflected ceiling;
   f. Electrical; and
   g. Elevations, sections, and details. (See Chapter 4.)

8. Various options in graphic representations or information displayed are available to the draftsperson, especially in the furniture plans and electrical plans. (See text.)

9. The existing conditions plan, indicating existing conditions, should include all architectural details and measurements. It is the basic plan, back to which a contractor will refer if any questions arise on subsequent plans. Existing condition elements should be in the heaviest line weight, since it is an existing conditions plan. All other notations, measurements, or indications should be in medium or guideline weight, depending on their function.

10. The demolition plan, indicating demolition, should indicate all measurements of those elements to be demolished, in addition to any reference measurement necessary to begin demolition. All elements to be demolished should be in heavy line weight in order to emphasize that it is a demolition plan. All other information and elements should be in either medium or guideline weights, depending on the requirements of the plan.

11. A construction plan, indicating construction, must indicate all measurements of those elements to be constructed in addition to any reference point necessary to begin construction at a specified point. Composition of construction (e.g., metal, wood, plaster, glass, etc.) should also be indicated via pouchering. All elements to be constructed should be in heavy line weight in order to emphasize that it is a construction plan. All other elements should be in medium or guideline weights, depending on the requirements of the plan.

12. A furniture plan, undressed or dressed, indicates placement of furniture. It may or may not include measurements or labeling of furniture. All furniture outlines should be in heavy line weight and all other elements in medium or guideline weight, depending on the requirements of the plan. Vary line weight of dressing methods to achieve a more effective drawing. Use overlays, underlays, stamps, or burnishing as may be desired.

13. A reflected ceiling plan is the plan of a ceiling as though it were transparent. It indicates placement of all ceiling luminaires, or luminaires above eye-level (e.g., sconces) in addition to ceiling molding, tiles, and HVAC vents. It must also give the contractor at least two measurements perpendicular to each other so that installation of the luminaire is at an exact spot that cannot be misinterpreted. All symbols for luminaires should be in heavy line weight to indicate that it is a reflected ceiling plan (overlapping structures and tiles may be put in medium line weight). All other elements should be in medium or guideline weight, depending upon the requirements of the plan. A notation of E.Q. (equidistant) may be used if luminaires are to be centered.

14. An electrical plan indicates to the contractor where outlets and switches are to be located, in addition to the circuitry of a project. Two measurements are required for proper specification. These are the measurement of installation away from an architectural element and the measurement A.F.F. (above finished floor) of installation. All electrical elements should be in heavy line weight to emphasize that this plan is an electrical plan. All other elements should be indicated in medium or guideline weights, depending on the requirements of the plan.

15. All plans should be clearly labeled as to type of plan.

16. Give as much written explanation as may be required to further explain graphic representations.

QUIZ QUESTIONS

1. When are borders used on drafting paper?

2. In what line weight should they be drafted?

3. What is a legend box and where is it placed? What information should be contained therein?

4. What is a title box, where is it placed, and what information should be contained therein?

5. List the various line weights used in both of the above and indicate for what section each weight is used.

6. Looking at something face-on would be considered what type of view?

7. What view is a floor plan considered to be?

8. Name the various types of plans and give a definition for each.

9. List the important points that must be covered in each of the types of plans.

10. What line weights are used for each of the aspects of those plans?

11. What types of plans may be combined if they are relatively simple?

12. What is the definition of: E.Q., A.F.F. and A.B.F.?

EXERCISES

1. Prepare ten sheets of drafting paper with border lines, and legend and title boxes.

2. Using one sheet per plan, redraft each of the plans illustrated in the text and use those measurements indicated on the plans for reference points. These plans should include:
   A. Existing Conditions
   B. Combination Existing Conditions
   C. Demolition
   D. Construction
   E. Combination Demolition/Construction
   F. Furniture Plan
   G. Reflected Ceiling Plan
   H. Electrical Plan
   I. Alternate Method of Electrical
   J. Combination Reflected Ceiling/Electrical.

   Make sure you understand what you are drafting and why you are drafting it. Be sure to have a full comprehension of each plan and the interrelationships of each of their measurements.

4
ELEVATIONS, SECTIONS, AND DETAILS

ELEVATIONS AND SECTIONS

Our eyes perceive objects as having height, width, and depth. However, the draftsperson, for the purpose of drafting elevations and sections, must learn to perceive objects as having only height and width. Until one is capable of doing so, difficulty in drafting and understanding elevations and sections may be encountered. Our eyes must be trained to think in terms of two dimensions only. An illusion of depth may be created only through the use of shading and degrees of intensity of line weights (the darker a line is, the closer to the viewer an object appears).

An elevation is a drawing of an object or structure as being projected vertically, without perspective, parallel to one of its sides. In other words, looking directly at an object while standing parallel to or in front of it. A section is the representation of an object as it would appear if it were cut by a plane showing its internal structure. That is to say, it is a somewhat x-ray view of the object while standing parallel to or in front of it. Elevations and sections are drafted to scale.

Elevations and sections differ from perspective drawings not only in their lack of depth, but in that they are drawn to scale. Our text is limited to only elevations and sections. Perspective drawings, for purposes of this book, are classified more as art work than as technical drafting.

The reason one drafts an elevation or section is to further clarify a design concept for either a client or a contractor. However, not all design concepts are best represented by elevations and sections. Usually, elevations and sections are used for the simplest and most uncluttered viewpoint. If depth, though, is an important aspect of the concept, then one should use a perspective drawing.

When a decision has been made to draft an elevation or section, this elevation should be keyed to the appropriate floor plan in order to orient the viewer properly. This key tells the viewer where he is standing (how far away from the elements to be shown in elevation) and how far to the left and right in the general floor plan he or she can see.

There are two types of elevations. One may draft a furniture elevation, which shows the furniture concept in a very simplified form; or an architectural elevation, which would show the architecture of a design concept. Either elevation may or may not need further clarification via measurement indication. The elevation's purpose and by whom it is to be used, such as client or contractor, would determine whether or not measurement indication would be required. If the elevation is for a furniture maker or building contractor, measurements would be required.

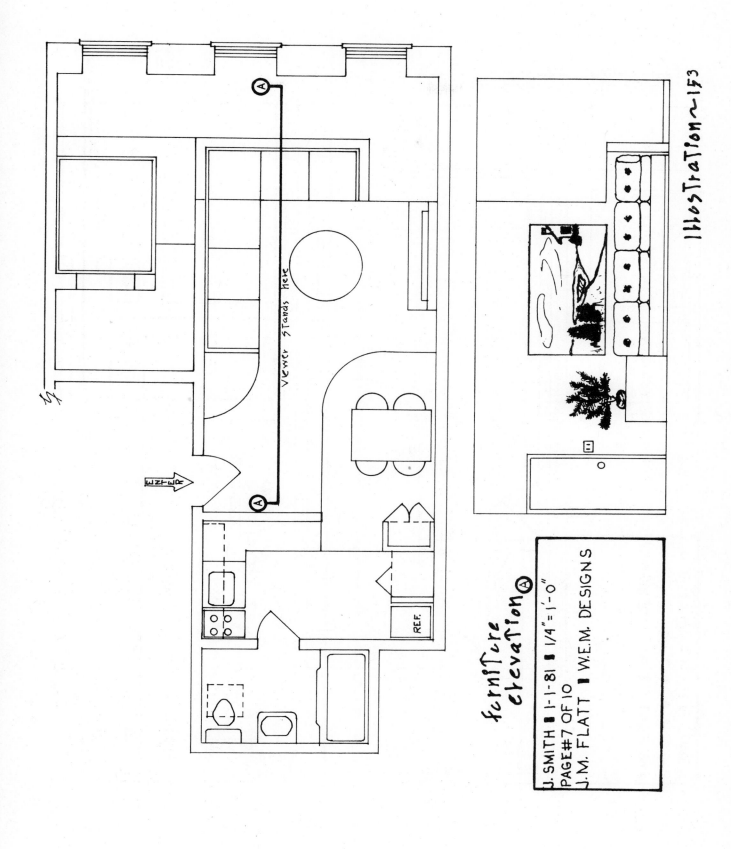

Viewer stands here

ENTER

Illustration ~ 153

furniture
elevation Ⓐ

J. SMITH ▪ 1-1-81 ▪ 1/4"=1'-0"
PAGE #7 OF 10
J.M. FLATT ▪ W.E.M. DESIGNS

REF.

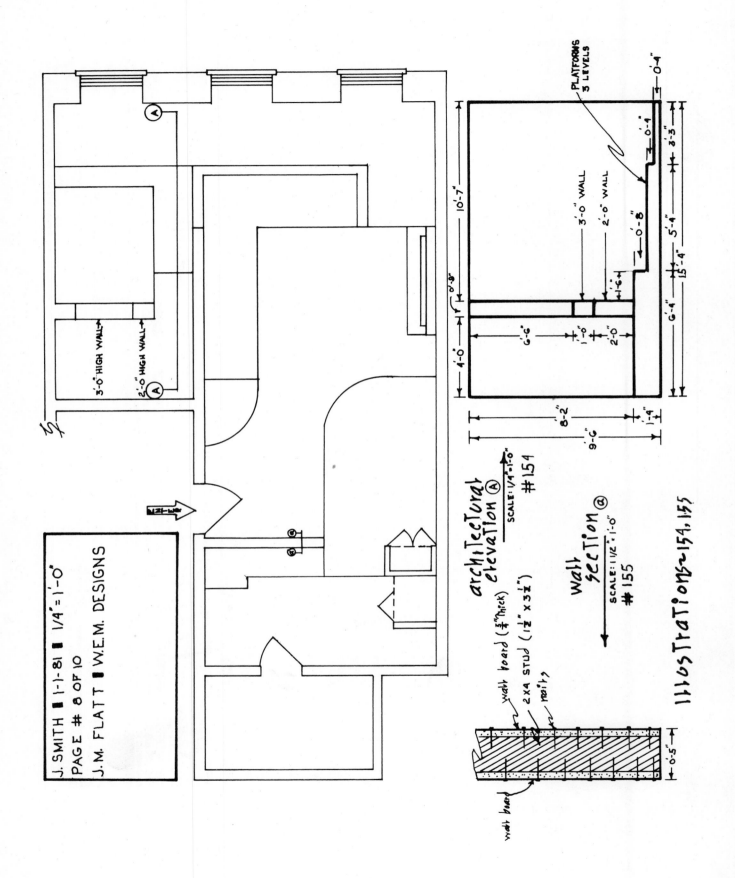

A

3'-6" HIGH WALL
2'-0" HIGH WALL

A

N24" Wide

architectural
elevation Ⓐ
SCALE: 1/4"=1'-0"
#154

PLATFORMS
3 LEVELS

3'-0" WALL
2'-0" WALL

10'-7"
4'-0"
1'-9"
6'-6"
1'-0"
2'-0"
6'-4"
15'-4"
5'-4"
3'-3"
0'-4"
0'-4"
0'-8"
1'-6"
8'-2"
1'-4"
9'-6"

wall
section Ⓐ
SCALE: 1 1/2"=1'-0"
#155

wall board (3/4" thick)
2x4 stud (1 1/2" x 3 1/2")
nails

wall board

0'-5"

Illustrations~154,155

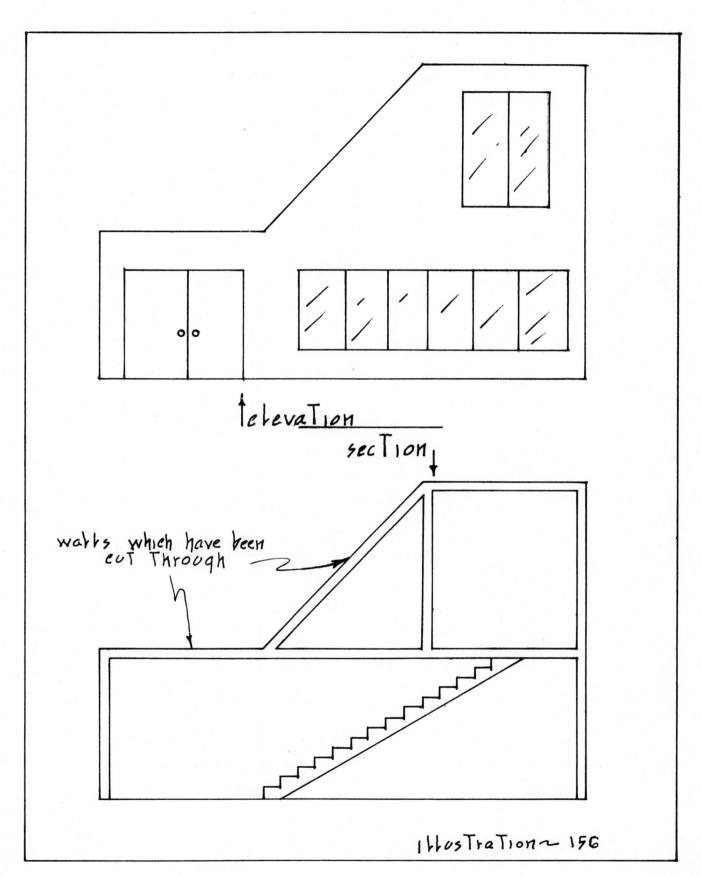

↑elevaTion

secTion↓

walls which have been
cuT Through

ILLusTraTion~ 156

107

ALL HANDLES ARE WIRE PULLS, 3" FROM CENTER OF LEFT MOUNTING TO CENTER OF RIGHT MOUNTING, FINISHED IN BRUSHED STAINLESS STEEL.

elevation

SCALE: 1/4" = 1'-0"

COUNTER TOP

EQ

EQ

DRAWERS

CABINETS

ADJUSTABLE SHELF

TOE SPACE

36"

1 1/2"

3 4"
6"

30 1/2"

24 1/2"

3"

9"

13"

13"

3 4"

1"

16"

6"

12"

16"

44"

EQ

EQ

H

detail drawing of area keyed in circle above.

1 FLAKE BOARD

1/16"
1/8"

3/4"

1" LAMINATE
16

1" FLAKE BOARD

J. SMITH ▌ 11-1-81 ▌ SCALE: 3"=1'-0"
PAGE #9 OF 10
J. M. FLATT ▌ W.E.M. DESIGNS

illustration~157

108

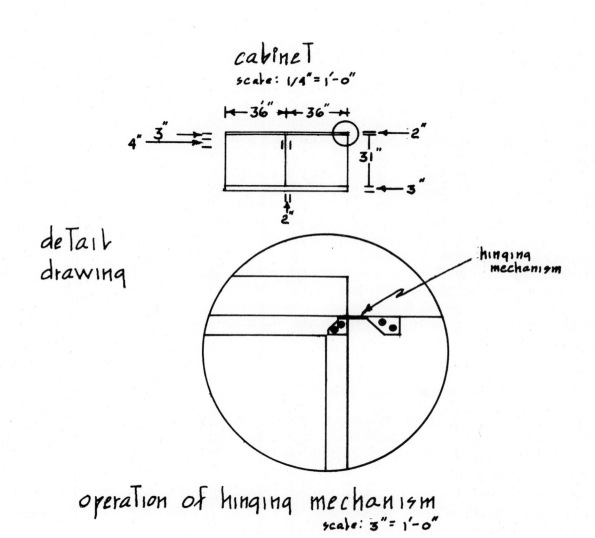

cabineT
scale: 1/4" = 1'-0"

deTail
drawing

operaTion of hinging mechanism
scale: 3" = 1'-0"

illusTraTion ~ 158

However, if the elevations are for a client's use, measurements would be somewhat optional.

Elevations and sections may either be placed on a separate sheet of paper, or drafted on the plan to which they pertain, depending upon logic and the draftsperson's preference. (See illustrations #153 to #156.)

DETAIL DRAWINGS

A detail drawing is employed to further explain a smaller element of a design concept, for example, the hinging mechanism on a piece of cabinetry or the construction of a window. (See illustration #157.) A detail drawing would be done to scale. However, this scale would be larger than the one used on the related plans; that is, instead of $\frac{1}{4}'' = 1'-0''$, a scale of $3'' = 1'-0''$ would possibly be used. A detail drawing must also be keyed to the original plan. However, a detailed drawing is keyed in a somewhat different manner than an elevation or section. In keying a detail drawing, one usually uses a large circle encompassing the element being detailed. (See illustrations #157 and #158.)

Detail drawings are especially useful when designing custom furniture. These drawings can indicate to the contractor or cabinetmaker the general feeling or look that the designer ultimately wants the element to have and how it is to operate. Detail drawings can also serve as another means of expressing to the client a particular aspect of the design concept.

REVIEW OF IMPORTANT POINTS

1. An elevation is a vertical, two-dimensional drawing which is drawn to scale and which has no depth. It is used to further clarify an idea. This should be done from the clearest point of view and keyed to the appropriate plan.

2. A section is an x-ray view of a structure and is also used to clarify an idea. It is drawn to scale and has no depth.

3. A detail drawing may be employed to emphasize a particular aspect of a design element, or to clarify, or indicate the method of construction. It too is drawn to scale, but may indicate depth for construction purposes. Usually the scale of the detail drawing is larger than that used for other plans.

4. A perspective drawing is not drawn to scale. It shows depth and is used to further explain a design concept.

5. Elevations, sections, or detail drawings may either be shown on separate plans or on the plan to which they relate. One should choose the method that seems most logical for the project.

QUIZ QUESTIONS

1. What is an elevation, and name two types?
2. What is a detail drawing?
3. What is a section drawing?
4. When is each of the above used?
5. Name three characteristics of each of the above drawings.
6. How does the scale of a detail drawing differ from that of a floor plan?
7. What dimension does an elevation and section drawing lack?
8. How does a perspective drawing differ from an elevation?

EXERCISES

1. Prepare four sheets of paper with border lines, etc.
2. Draft:
   A. Architectural elevation shown in the text
   B. Furniture elevation shown in the text
   C. Section shown in the text
   D. Detail drawing shown in the text.
   Use ½″ = 1′–0″ scale for each drawing.

5
DRESSING A PLAN

A dressed plan gives the client a fuller picture of the design concept by adding more visual details and elements to the drawing. Usually, the plan that is dressed is the furniture plan. A dressed furniture plan gives the client a better picture of how the room or rooms will look when fully redesigned and furnished.

To dress a plan, one has several techniques at his or her disposal, of which one or all of the following may be used:

1. Shading of various levels of flooring, e.g., platforms;
2. Conveyance of tufting or upholstery on furniture;
3. Indication of construction material, e.g., mirror, glass (see coffee table, illustration #159);
4. Indication of fabric texture, design, or color;
5. Placement of rugs or use of fringes on rugs;
6. Indication of pillows;
7. Trees;
8. Indication of a person (for elevations and sections) to give relationships of size;
9. Indication of curves (for elevations and sections);
10. Various floor patterns, e.g., veining of marble, indication of tile or carpeting;
11. Pressure sensitive films (letra set or prestype for indicating unusual patterns, letterings, or numberings). These are like contact paper and require skill and practice before use on a plan. (See Chapter 1, *Tools of the Trade*.)
12. Indication of fixtures, e.g., faucet, showers;
13. Indication of appliances, e.g., burners on a stove;
14. Indication of use, e.g., a turned-down sheet on a bed;
15. Changing of leads by use of a combination of soft leads and hard leads to achieve differing line effects and weights;
16. Stamps, and
17. Burnish plates.

Compare illustration #146 (non-dressed) with illustration #159 (dressed). The difference between the two is quite evident, so it would follow that the general concept of the design solution portrayed by each

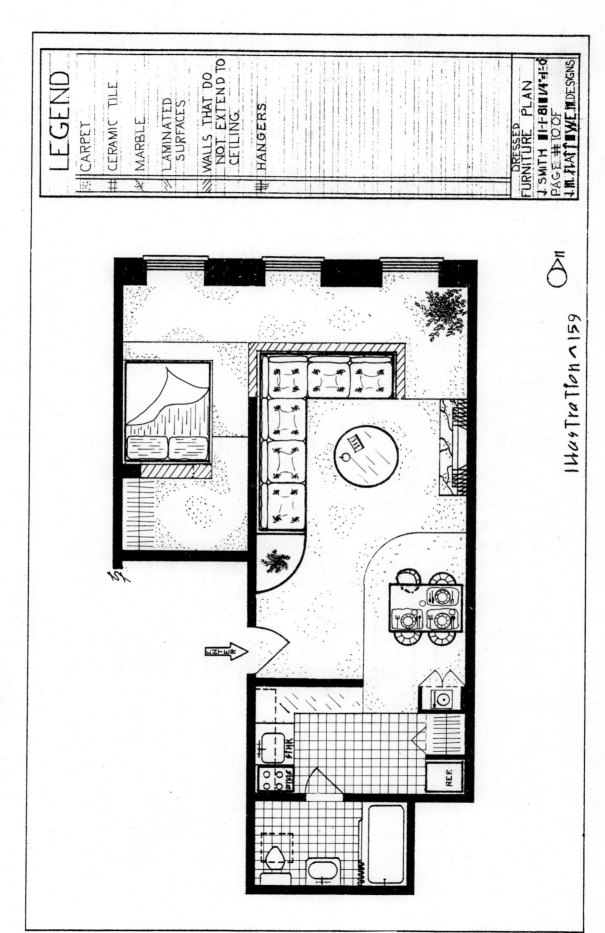

would also be vastly different. One must determine the requirements of a project and draft the type plan accordingly.

Dressing a furniture plan requires time, patience, and some artistic capability. Several practice sessions should be used before dressing the final plan so that the desired effects can be achieved with confidence for success. One should not try to "overdress" a plan by trying to indicate everything possible. Some aspects, those determined by the judgment and preference of the draftsperson, should be left to the viewer's imagination. Usually, they are those aspects of a project which may cause confusion for the viewer if indicated graphically. Each project will require a different type and extent of dressing.

QUIZ QUESTIONS

1. What is meant by dressing a plan?
2. When would one dress a plan?
3. Name seventeen methods of dressing a plan and a brief description of each.
4. What plan is usually dressed?
5. What dictates the extent of dressing a plan?
6. Is the statement "the more a plan is dressed the better" true? Why?

EXERCISES

1. Prepare several sheets of paper with borders.
2. Draft the dressed furniture plan shown in the text.
3. Review those alternatives available to you when dressing a plan and dress this plan using at least ten of the alternatives.
4. Repeat this process using five different alternatives.

6
REPRODUCTION OF PLANS

REASONS FOR REPRODUCTION

There are many reasons why one may wish to have plans reproduced. First of all, it is always wise to have more than one copy of any document. In this way, if the original should be destroyed or lost, one may always turn to the reproduction. It would be most disastrous if no copies were available of an important plan should the original not be available. Many hours of measuring, redrafting, and thinking time will be saved simply by having reproductions made in one form or another. In addition, one may need to distribute copies of a plan to many people. Such would be the case with contractors. One may wish several contractors to put out a bid or estimate on executing the work. So that all contractors can bid at the same time, one can simply have several copies made of a plan and distribute them simultaneously. Thirdly, when finally on a job, it is advisable that each type of contractor, such as construction, demolition, electrical, etc., have his or her own copy of the plan. In this way, various steps of the project may be in progress at the same time. Each contractor would not have to rely on another contractor's copy of the plan.

TYPES OF REPRODUCTIONS

There are several types of reproduction from which one may choose. The type one decides upon would be dictated by the functional and aesthetic needs of the project. One may wish to make corrections on a plan and therefore would want a type of reproduction on which corrections may be made. One may be giving a presentation of a furniture plan to a client and prefer to have that plan relate color-wise or style-wise to the design concept. In that case, it would be best to have a plan reproduced in one of the limited colors or finishes available that would so relate. The following are types of reproductions available to the draftsperson:

1. *Blue Line Print:* All lines drafted appear in blue ink, the background is white.
2. *Black Line Print:* All lines drafted appear in black ink, the background is white.
3. *Brown Line Print:* All lines drafted appear in brown ink, the background is white.
4. *Sepia Print:* Similar to brown line print, but permits corrections to take place via an erasing correction fluid.

5. *Photographic Print:* A photograph of a print and it would be used for client presentations, not contractors' purposes. It gives a more finished look.

6. *Photostatic Print:* A somewhat xerox version of a print.

7. *Negative Print:* A reverse print. What was drafted appears in white and the background is black. Excellent for very dramatic presentations.

Costs

Costs can range from very inexpensive (under $1.00) for blue, brown, and black line prints to intermediate cost (under $10.00) for sepia, photographic, and photostatic copies to expensive (about $30.00–$60.00) for negative prints. Costs will also vary according to the size of the print.

One should first decide the project needs, in addition to taking into consideration the pros and cons of each plan, and then proceed to have the proper reproductions made.

Reproduction Clarity

Plans usually reproduce better when the original is drafted on a reasonably clear paper, and in ink. However, careful draftsmanship with the use of lead will provide a drawing that will reproduce clearly, just as clearly as though it were drafted in ink.

REVIEW OF IMPORTANT POINTS

1. Basically, there are seven types of prints. They are:
   a. Blue Line;
   b. Black Line;
   c. Brown Line;
   d. Sepia
   e. Photostatic;
   f. Photographic; and
   g. Negative.

2. Type and amount of prints used would be dictated by project requirements.

3. Prints reproduce better when drafted in ink.

4. Costs of prints vary according to size and type of print requested. Prices range from under $1.00 for blue, black, and brown line prints to $3.00 through $60.00 for sepia, photostatic, photographic, and negative prints.

5. Multiples of prints are useful when copies of the same print must be distributed to more than one workman or contractor for bidding or use on the job.

QUIZ QUESTIONS

1. What is a blue line print?
2. What is a black line print?
3. What is a brown line print?
4. What is a sepia print?
5. What is a photostatic print?
6. What is a photographic print?
7. What is a negative print?
8. When is each of the above used?
9. Give approximate prices for each of the above types of prints.

EXERCISES

Draft several types of plans and have each reproduced in the methods mentioned in this chapter. This will demonstrate how your drafting skills reproduce and what adjustments in your techniques must be made, if any.

7
USEFUL OVERALL EXERCISES

The following is a list of exercises that will promote the advancement of drafting skills.

A. CHARACTERISTICS OF DIFFERENT LEADS, LINE WEIGHTS, AND MEASUREMENT INDICATION

Using ¼" scale, draft thirty lines of different lengths and weights (guide, medium, and heavy). Indicate the measurement of each line in feet and inches using the seven methods illustrated in the text.

B. LETTERING AND NUMBERING

On an 8½" × 11" sheet of blank paper, draft horizontal guidelines, alternating the space between these lines at ⅛" and ³⁄₁₆" intervals. Print an article from the newspaper, using the ³⁄₁₆" spaces as the writing surface and the ⅛" spaces as spaces. Repeat this process for ten to fifteen sheets; allow sufficient time for your hand to rest, and substitute ⅛" measurement for ³⁄₁₆" measurement.

C. PLAN DRAFTING

1. Draft an existing conditions plan of your apartment or house. Poucher each wall according to the composition of its materials. Key the pouchering technique to the legend as one would do as a matter of course for all plans.

2. Draft a demolition plan as though you were instructing a contractor to remove some of the walls, doors, and windows.

3. Draft a construction plan, indicating to the contractor (a) where the new walls, doors, and windows are to be placed, (b) what their composition is, and (c) draft the appropriate schedules.

4. Draft a furniture plan that would be appropriate for the new space that you have created. After drafting a "non-dressed" plan, draft one that would be considered "dressed."

5. Draft a reflected ceiling plan to correspond with the new furniture.

6. Draft an electrical plan related to the reflected ceiling plan.

7. Draft two simple elevations and one section drawing.

8. Draft a detail drawing of the corner of a bookcase or a door with a hinge using 3" = 1'–0" as a scale.

Note: (a) provide legend, title boxes, and finish schedules where applicable; and

(b) be sure to give all necessary measurements and information required for each respective plan. (Refer to Chapter 4, *The Various Types of Plans.*)

D. REPRODUCTION OF PLANS

Have the following types of prints made of all your drawings so that you can see how each reacts to your drafting pressure and style: blue line, black line, brown line, sepia, photostatic, photographic, and negative.

CONCLUSION

So you're a draftsperson now! Well, not quite. Remember, the material covered in this book, at the most, is considered elementary. It does, however, set the groundwork for a discipline—a discipline that by its very nature, and more so of necessity, is considered to be, and must remain, artistically scientific.

Complete digestion of the material in this text may only be accomplished through painstaking hours of tedious practice. For the fullest comprehension of the material covered, It is strongly suggested that one reread each chapter until all facets set forth are fully assimilated. In doing so step-by-step, one will have provided an excellent basis on which a higher degree of drafting expertise may be built.

In all probability, it will also be found that the regimented, exacting manner of measuring, describing, explaining, and portraying graphically or by written word a physical concept will be of great use in the many other areas of our life. As with all else in life, perfection may only be achieved through practice. However, the ability to supply the means of transference from concept to reality is a most rewarding talent. The challanges with which we are supplied by our minds should and can be fulfilled.

Happy drafting.

APPENDIX

Standard Furniture, Cabinet, and Appliance Sizes

BEDROOM

Bed Type	Length	x	Width
Bunk	75″	x	30″
Dormitory	75″	x	33″
	80″	x	36″
Double	75″	x	54″
	80″	x	54″
	84″	x	54″
King	80″	x	72″
	80″	x	76″
	84″	x	72″
	84″	x	76″
Queen	80″	x	60″
	84″	x	60″
Three-Quarters	75″	x	48″
	80″	x	48″

Bed Type	Length	x	Width
Twin	75″	x	39″
	80″	x	39″
	84″	x	39″
Sofa Bed	87″	x	31″
	91″	x	32″
	79″	x	34″

Night Tables	Length	x	Width	x	Height
	24″	x	15″	x	22″
	22″	x	16″	x	22″
	24″	x	18″	x	22″
	22″	x	22″	x	22″

Chest of Drawers	Length	x	Depth	x	Height
	20″	x	16″	x	50″
	26″	x	16″	x	31″
	28″	x	15″	x	34″
	32″	x	17″	x	43″
	36″	x	18″	x	45″

Desk	Length	x	Depth	x	Height
	33″	x	16″	x	29″
	36″	x	16″	x	29″
	40″	x	20″	x	30″
	43″	x	16″	x	30″

Dresser Type					
Double	48″	x	18″	x	30″
	50″	x	18″	x	30″
Triple	52″	x	16″	x	30″
	60″	x	18″	x	30″

Wardrobe					
	36″	x	22″	x	66″
	48″	x	22″	x	66″
	60″	x	22″	x	66″

BATHROOM

Bathtub Type	Width	x	Length	x	Height
Rectangular	30–¾″	x	54″	x	16″
	30″	x	60″	x	14″

Bathtub Type	Width	x	Length	x	Height
Rectangular	30″	x	60″	x	16–½″
	31″	x	60″	x	15–½″
	31–½″	x	60″	x	16″
	31–½″	x	66″	x	18″
	30–¾″	x	72″	x	16″
Square	37″	x	42″	x	12″
	42″	x	48″	x	14″

Water Closet Type	Width	x	Length	x	Height
Floor Mounted Two-Piece	17″	x	25–½″	x	29–½″
	21″	x	26–¾″	x	28″
	21″	x	28–¾″	x	28″
Floor Mounted One-Piece	20–⅜″	x	27–¾″	x	20″
	20–⅜″	x	29–¾″	x	20″
Wall Hung Two-Piece	22–½″	x	26″	x	31″
Wall Hung One-Piece	14″	x	24–¼″	x	15″

Wall Hung Sink	Width	x	Depth
	19″	x	17″
	20″	x	18″
	22″	x	19″
	24″	x	20″

Medicine Cabinet Type	Width	x	Depth	x	Height	
One-Shelf	36″	x	12″	x	12″	(Two doors)
	39″	x	12″	x	12″	(Two doors)
	42″	x	12″	x	12″	(Two doors)
	30″	x	12″	x	15″	(Two doors)
	33″	x	12″	x	15″	(Two doors)
	36″	x	12″	x	15″	(Two doors)
	39″	x	12″	x	15″	(Two doors)
	42″	x	12″	x	15″	(Two doors)
Two-Shelf	24″	x	12″	x	18″	(Two doors)
	27″	x	12″	x	18″	(Two doors)
	30″	x	12″	x	18″	(Two doors)
	33″	x	12″	x	18″	(Two doors)
	36″	x	12″	x	18″	(Two doors)
	39″	x	12″	x	18″	(Two doors)

Medicine Cabinet Type	Width	x	Depth	x	Height	
Two-Shelf	42″	x	12″	x	18″	(Two doors)
	15″	x	12″	x	24″	(One door)
	18″	x	12″	x	24″	(One door)
	21″	x	12″	x	24″	(One door)
	24″	x	12″	x	24″	(Two doors)
	27″	x	12″	x	24″	(Two doors)
	30″	x	12″	x	24″	(Two doors)
	33″	x	12″	x	24″	(Two doors)
	36″	x	12″	x	24″	(Two doors)
	39″	x	12″	x	24″	(Two doors)
	42″	x	12″	x	24″	(Two doors)
Three-Shelf	12″	x	12″	x	30″	(One door)
	15″	x	12″	x	30″	(One door)
	18″	x	12″	x	30″	(One door)
	21″	x	12″	x	30″	(One door)
	24″	x	12″	x	30″	(Two door)
	27″	x	12″	x	30″	(Two door)
	30″	x	12″	x	30″	(Two door)
	33″	x	12″	x	30″	(Two door)
	36″	x	12″	x	30″	(Two door)

Medicine Cabinet Type	Width	x	Depth	x	Height	
Three-Shelf	39″	x	12″	x	30″	(Two door)
	42″	x	12″	x	30″	(Two door)

Base Cabinet Type	Width	x	Depth	x	Height	
Drawer and Door with Shelf	9″	x	24″	x	34½″	(One door)
	12″	x	24″	x	34½″	(One door)
	15″	x	24″	x	34½″	(One door)
	21″	x	24″	x	34½″	(One door)
	24″	x	24″	x	34½″	(2 door/2 drawer)

Base Cabinet Type	Width	x	Depth	x	Height	
	27″	x	24″	x	34½″	(2 door/2 drawer)
	30″	x	24″	x	34½″	(2 door/2 drawer)
	33″	x	24″	x	34½″	(2 door/2 drawer)
	36″	x	24″	x	34½″	(2 door/2 drawer)
	39″	x	24″	x	34½″	(2 door/2 drawer)
	42″	x	24″	x	34½″	(2 door/2 drawer)
	45″	x	24″	x	34½″	(2 door/2 drawer)
	48″	x	24″	x	34½″	(2 door/2 drawer)
Sink Front Base	21″	x	24″	x	34½″	(One door)
	24″	x	24″	x	34½″	(Two doors)
Sink Front Base	27″	x	24″	x	34½″	(Two door)
	30″	x	24″	x	34½″	(Two door)
	33″	x	24″	x	34½″	(Two door)
	36″	x	24″	x	34½″	(Two door)
	39″	x	24″	x	34½″	(Two door)
	42″	x	24″	x	34½″	(Two door)
	45″	x	24″	x	34½″	(Two door)
	48″	x	24″	x	34½″	(Two door)
Three Drawer Base	15″	x	24″	x	34½″	
	18″	x	24″	x	34½″	
	21″	x	24″	x	34½″	
	24″	x	24″	x	34½″	
Five Drawer Base	27″	x	24″	x	34½″	
	30″	x	24″	x	34½″	

KITCHEN

Refrigerator Type	Width	x	Height	x	Depth
9 cubic feet	24″	x	56″	x	29″
12 cubic feet	30″	x	68″	x	30″
14 cubic feet	31″	x	63″	x	24″

Refrigerator Type	Width	x	Height	x	Depth
19 cubic feet	34″	x	70″	x	29″
21 cubic feet	36″	x	66″	x	29″

Stove Type	Width	x	Height	x	Depth
Standard Free-Standing Range	20″	x	30″	x	24″
	21″	x	36″	x	25″
	30″	x	36″	x	26″
	40″	x	36″	x	27″
Double Oven Range	30″	x	61″	x	26″
	30″	x	64″	x	26″

Stove Type	Width	x	Height	x	Depth
	30″	x	67″	x	27″
	30″	x	71″	x	27″
Drop-In Range	23″	x	23″	x	22″
	24″	x	23″	x	22″
	30″	x	24″	x	25″
Built-In Cook Top	12″	x	2″	x	18″
	24″	x	3″	x	22″
	48″	x	3″	x	22″

Sink Type	Width	x	Depth
Double Compartment	32″	x	21″
	36″	x	20″
	42″	x	21″
Single Compartment	24″	x	21″
	30″	x	20″

DINING ROOM

Dining Table Type	Length	x	Width	x	Height
Rectangular	42″	x	30″	x	29″
	48″	x	30″	x	29″
	48″	x	42″	x	29″
	60″	x	40″	x	28″
	60″	x	42″	x	29″
	72″	x	36″	x	29″
Oval	54″	x	42″	x	28″
	60″	x	42″	x	28″
	72″	x	40″	x	28″
	72″	x	48″	x	28″
	84″	x	42″	x	28″

Dining Table Type	Diameter	x	Height
Round	32″	x	28″
	36″	x	28″
	42″	x	28″
	48″	x	28″

China Cabinet or Hutch	Length	x	Width	x	Height
	48″	x	16″	x	65″
	50″	x	20″	x	60″
	62″	x	16″	x	66″
Buffet	36″	x	16″	x	31″
	48″	x	16″	x	31″
	52″	x	18″	x	31″

Dining Chairs (Seat Ht: 16″)	Width	x	Depth	x	Height
	17″	x	19″	x	29″
	20″	x	17″	x	36″
	22″	x	19″	x	29″
	24″	x	21″	x	31″

LIVING ROOM	Width	x	Depth	x	Height
Sofa With Arms	72″	x	36″	x	28″
	76″	x	35″	x	35″
	84″	x	36″	x	37″
	87″	x	31″	x	31″
	88″	x	32″	x	29″
	91″	x	32″	x	30″
Sofa With No Arms	72″	x	30″	x	30″
	74″	x	30″	x	30″
	90″	x	30″	x	30″
Desk	50″	x	21″	x	30″
	50″	x	22″	x	30″
	55″	x	26″	x	29″
	60″	x	30″	x	29″
	72″	x	36″	x	29″
Chair Type Lounge	28″	x	32″	x	29″
	33″	x	31″	x	31″
	33″	x	32″	x	30″

Chair Type	Width	x	Depth	x	Height
Lounge	34″	x	36″	x	37″
	35″	x	35″	x	35″
Recliner	31″	x	30″	x	40″
	32″	x	34″	x	40″
	32″	x	35″	x	41″
	36″	x	37″	x	41″
Small Arm	18″	x	18″	x	29″
	21″	x	22″	x	32″
Table Type Cocktail	35″	x	19″	x	17″
	50″	x	18″	x	15″
	54″	x	20″	x	15″
	56″	x	21″	x	16″
	57″	x	19″	x	15″
	58″	x	20″	x	15″
	61″	x	21″	x	17″
	66″	x	20″	x	15″

Table Type	Width	x	Depth	x	Height
Corner	28″	x	28″	x	20″
	30″	x	30″	x	15″
	36″	x	36″	x	16″
End	21″	x	28″	x	20″
	22″	x	28″	x	21″
	26″	x	20″	x	21″
	27″	x	19″	x	22″
	28″	x	28″	x	20″

Round Cocktail	Diameter	x	Height
	24″	x	16″
	30″	x	15″
	36″	x	16″
	42″	x	15″
	48″	x	16″

Bunching	Width	x	Depth	x	Height
	20″	x	20″	x	15″
	19″	x	19″	x	15″
	21″	x	21″	x	16″

Ottoman	Width	x	Depth	x	Height
	22″	x	18″	x	13″
	22″	x	22″	x	16″
	24″	x	19″	x	16″
Love Seat	47″	x	28″	x	36″
	54″	x	30″	x	36″
	59″	x	36″	x	37″
Shelf Units	17″	x	10″	x	60″
	24″	x	10″	x	60″
	36″	x	10″	x	36″
	36″	x	10″	x	60″
	48″	x	10″	x	60″

Approximate Standard Window Sizes

DOUBLE HUNG	WIDTH	x	HEIGHT
	2′–6″	x	3′–2″
	2′–10″	x	3′–2″
	3′–2″	x	3′–2″
	2′–2″	x	3′–6″
	2′–6″	x	3′–6″
	2′–10″	x	3′–6″
	3′–2″	x	3′–6″
	3′–6″	x	3′–6″

DOUBLE HUNG	**WIDTH**	**x**	**HEIGHT**
	2'–2"	x	4'–2"
	2'–6"	x	4'–2"
	2'–10"	x	4'–2"
	3'–2"	x	4'–2"
	3'–6"	x	4'–2"
	2'–2"	x	4'–6"
	2'–6"	x	4'–6"
	2'–10"	x	4'–6"
	3'–2"	x	4'–6"
	3'–6"	x	4'–6"
	1'–9"	x	4'–10"
	2'–2"	x	4'–10"
	2'–6"	x	4'–10"
	2'–10"	x	4'–10"
	3'–2"	x	4'–10"
	3'–6"	x	4'–10"
	3'–10"	x	4'–10"
	2'–2"	x	5'–6"
	2'–6"	x	5'–6"
	2'–10"	x	5'–6"
	3'–2"	x	5'–6"
	3'–6"	x	5'–6"
	3'–10"	x	5'–6"
	2'–10"	x	5'–10"
	3'–2"	x	5'–10"
	3'–6"	x	5'–10"
	3'–10"	x	5'–10"
	3'–6"	x	6'–6"
	3'–10"	x	6'–6"

CASEMENT	**WIDTH**	**x**	**HEIGHT**
	2'–0"	x	3'–0"
	4'–0"	x	3'–0"
	6'–0"	x	3'–0"
	2'–0"	x	3'–5"
	4'–0"	x	3'–5"
	6'–0"	x	3'–5"
	2'–0"	x	4'–0"
	4'–0"	x	4'–0"
	6'–0"	x	4'–0"

2'–0"	x	5'–0"
4'–0"	x	5'–0"
6'–0"	x	5'–0"
2'–0"	x	6'–0"
4'–0"	x	6'–0"

HORIZONTAL SLIDING

3'–8"	x	2'–11"
4'–8"	x	2'–11"
5'–8"	x	2'–11"

HORIZONTAL SLIDING

WIDTH	x	HEIGHT
3'–8"	x	3'–7"
4'–8"	x	3'–7"
5'–8"	x	3'–7"
3'–8"	x	4'–3"
4'–8"	x	4'–3"
5'–8"	x	4'–3"
3'–8"	x	4'–11"
4'–8"	x	4'–11"
5'–8"	x	4'–11"
4'–8"	x	5'–7"
4'–8"	x	5'–7"
5'–8"	x	6'–3"

AWNING

WIDTH	x	HEIGHT	
3'–0"	x	2'–0"	Single Awning
4'–0"	x	2'–0"	
3'–0"	x	4'–0"	Double Awning
4'–0"	x	4'–0"	

AWNING

WIDTH	x	HEIGHT	
3'–0"	x	6'–0"	Stationary Center w/
4'–0"	x	6'–0"	Awning Top & Bottom
3'–0"	x	6'–0"	Stationary Top w/Awning
4'–0"	x	6'–0"	on Bottom

HOPPER

2'-9"	x	1'-4"
2'-9"	x	1'-8"
2'-9"	x	2'-0"

PICTURE

45"	x	37"
70"	x	37"
94"	x	37"
45"	x	50"
70"	x	50"
94"	x	50"
45"	x	60"
70"	x	60"
94"	x	60"

PICTURE

WIDTH	x	HEIGHT
45"	x	74"
70"	x	74"
94"	x	74"

Standard Door Sizes

INTERIOR

WIDTH	x	HEIGHT
2'-0"	x	6'-8"
2'-6"	x	6'-8"
3'-0"	x	6'-8"

EXTERIOR

3'-0"	x	6'-10"
3'-6"	x	6'-10"

GLOSSARY OF COMMON TERMS

A.B.F. Abbreviation for "above base floor."

Accordion Door. Doors that create a multiple of folds against each other when opened.

A.F.F. Abbreviation for "above finished floor."

Alcove. An indented opening off a wall of a larger area. Sometimes converted to storage, sleeping, or eating facility.

Apron. Horizontal trim under the stool of the interior window.

Architect's Scale Ruler. A ruler that converts measurement to a smaller representation of space, e.g., $\frac{1}{4}'' = 1'-0''$, $\frac{1}{8}'' = 1'-0''$, etc.

Archway. Opening in a wall, comprised of two sides and an overhead beam, which leads to another area.

Atrium. An open or glass enclosed hall, usually centrally located within a structure.

Attic. Space between a ceiling level and the roof.

Awning Window. Window with top hinging and end opening outward.

Axonometric Drawing. Also known as a paraline drawing. Drawing types include isometric, elevation, oblique, and plan oblique; it is a single-view drawing in which all lines are parallel and projected at an angle.

Balcony. A deck above ground level, projecting from either an interior or exterior wall.

Baluster. Small vertical member in a railing used between a top rail and the stair treads or a bottom rail.

Banister. A handrail with supporting posts used alongside a stairway.

Baseboard. A finishing board of any material (e.g., plastic, rubber, wood, marble) that covers the joining area of the wall and floor.

Bay Window. Any window with a projection away from the exterior of a building.

Beam. A horizontal structural element supporting a load.

Bifold Door. Doors that give the appearance of one side folding against the other.

Birch. A hardwood, light reddish-brown in color. Excellent for use in all cabinet work.

Black Line Print. A black and white print where background appears white and any lead or ink markings appear black.

Blue Line Print. A blue and white print where background appears white and any lead or ink markings appear blue.

Brick. A masonry unit composed of clay and shale. Usually rectangular in shape.

Brown Line Print. A brown and white print where background appears white and any lead or ink markings appear brown.

Casement Window. General term for a hinged window that opens outward. Usually made of metal.

Column. A vertical structural member supporting a load.

Combination Window. Window with one section (usually the top) of stationary glass and the other either a hopper or awning window.

Compass. Instrument used to draft circles or parts thereof.

Concrete. A mixture of sand, gravel, cement, and water to produce a masonry compound for construction.

Cornice. A horizontal structure projecting from the wall at ceiling level.

Cone Molding. Concave molding used in cabinet and built-in construction.

Crown Molding. Molding used at the top of cabinetry and where ceilings meet walls.

Curtain Wall. An exterior wall providing no structural function for support.

Dado Joint. A joining construction method in cabinetry where one board has a groove cut to receive another board for a tight and secure fit.

Demolition Plan. A drafted drawing indicating which elements of a project are to be eliminated.

Design Concept. An idea or solution of a decorative nature.

Detail Drawing. A drawing executed to scale that describes in detail a specific aspect of a design concept.

Door Jamb. Two vertical elements held together by a horizontal element forming the inside of a door opening.

Door Stop. Members on the door jambs against which the door closes.

Double Action Door. Door that swings both inward and outward.

Double Glaze Window. One pane of glass comprised of two sheets of glass with insulation provided by airspace between the two sheets.

Double Hung Window. A vertically sliding window composed of a top and bottom window, each capable of independent movement.

Double L Stairs. Stairs that form two "L"-shapes and have a landing between each.

Douglas Fir. A softwood ranging in color from yellow to pale red. Usually converted into plywood.

Drafting Board. A small horizontal surface, which may or may not incline, used specifically for drafting.

Drafting Brush. A brush used to discard erasures from a drawing.

Drafting Table. A large horizontal surface with leg supports to the floor, which may or may not incline, used specifically for drafting.

Dressing. The artistic embellishment of a drawing used to further clarify an element of a design concept, e.g., indication of tufting, shading, etc.

Dry Construction. Interior covering materials, such as wallboard or gypsum, which is applied in large panels to form a wall.

Duplex Outlet. Electrical receptacle capable of receiving two plugs.

Dutch Door. Door that has a top portion closing independently from the bottom portion closing.

Eastern Fir. A softwood used for minor construction, such as molding.

Electrical Plan. A drafted drawing executed to scale, indicating the circuity and location of all electrical elements, such as switches and outlets.

Elevation. A drafted drawing without depth executed to scale viewed from a head-on position.

Elevation Oblique. A projection of an elevation drafted at varying angles, thereby giving the viewer a depth perception. All vertical lines remain vertical and all parallel lines remain parallel.

Erasing Shield. A protective device used to cover any part of a line which is not to be erased, while exposing that part of a line that is to be erased.

Ergonometrics. The study of human body movement and its relationship to the space in which it normally functions.

Existing Conditions Plan. A drafted drawing executed to scale that indicates those conditions present in a project prior to any demolition or construction work being performed.

Face Brick. Better quality brick used for visible construction.

Fenestration. The pattern formed, or the placement of, windows in a wall.

Fire Brick. An especially heat-resistant brick that is used in fireplaces.

Flagstone. Flat stone used for walkways, walls, and steps.

Flush Door. Door that closes flush with adjoining walls.

French Curve. An instrument used to form irregular curves.

French Door. Two vertical doors closing against each other within a frame.

Furring. Construction method used to level the surface of a wall and prepare it for receipt of insulation or application of a topical treatment.

Ganging. The term used to indicate installation of several outlets or switches next to each other, thereby forming a row.

Guideline Weight. An exceptionally light line used for guiding lettering, numbering, and used as a basis for darker lines to be drawn at a later time.

Hardwood. Wood produced from broad-leaved trees or those trees that lose their leaves, e.g., birch, maple, oak, and walnut.

Hearth. Inner or outer floor surface of a fireplace, usually composed of brick, tile, or stone.

Heavy Line Weight. A dark line used to indicate those aspects of a drawing that are of primary importance.

Hickory. A hardwood ranging in color from brown to reddish brown. It may be used either as a veneer or as a construction material.

Horizontal Sliding Window. Window in which glass panes slide horizontally from side to side.

Hopper Window. Window hinged at bottom and opens outward.

H.V.A.C. Abbreviation for heating, ventilation, air conditioning units.

Insulation. Any material used for obstructing the transference of cold, heat, or sound from one area to another.

Interior Trim. General construction term used to denote all interior moldings and baseboards.

Isometric Drawing. A drawing drafted at a 30° angle from the horizontal and giving equal emphasis to all visible surfaces. All vertical lines remain vertical and all parallel lines remain parallel.

Jalousie Window. A window composed of long, narrow, horizontal panes of hinged glass.

Jamb. The side and top lining of a doorway, window, or other structural element.

Joist. A horizontal structural member supporting either a floor or ceiling structure.

Junction Box. The point at which the primary electrical source meets the on outlet or connection for an electrical fixture.

Lamp. The correct term for light bulb.

Landing. A platform between or at either end of stairs.

Leader. An arrow, pointing toward an element of a drafted drawing, giving a description or further explanation.

Leadholder. A mechanical device, in the shape of a pencil, that holds varying widths of lead.

Legend Box. That portion of a drawing that lists any symbol used in the drawing along with its definition.

Lens. A light diffusing element that has no openings in its surface.

Letra-Set. Manufacturer's name for plastic film sheets capable of producing numbers, letters, or graphic patterns that can be applied to a drawing surface by pressure.

Line Weight. The intensity of a line.

Louvre. A light-diffusing element that has openings in its surface.

L Stairs. Stairs that form the shape of an "L" with a landing between the stairs.

Luminaire. The correct term for a lighting fixture.

Mantel. The shelf above, and trim around, the opening of a fireplace.

Maple. A hardwood, usually light tan in color. It is used in quality construction where hardness is a primary consideration.

Masonry. Any building material such as brick, stone, concrete, gypsum, etc. that is bonded together to form a construction element, such as a wall.

Mechanical Pencil. A mechanical device, in the shape of a pencil, capable of holding very thin lead.

Medium Line Weight. A line of medium intensity used to indicate those elements of a drawing that are of secondary importance.

Miter Joint. Construction method used in cabinetry that employs two pieces of lumber cut at 45° angles and fitted together.

Mortise Joint. A construction method used in cabinetry to join two pieces of wood—one piece with a slot, usually cut edge-wise, made to receive the other piece of wood.

Mintin. Small construction element, either horizontal or vertical, used to connect glass panes in a window or door.

Mullion. The vertical member between two panes of glass in a window.

Mylar Film. A plastic transparent film-like paper used for drafting.

Negative Print. A black and white print in which the background is black and any ink or lead markings are white.

O.C. Abbreviation for location of a measurement indicating on center.

One Point Perspective. A drawing which depicts depth by having parallel lines converge to one vanishing point, thus creating the illusion or depth.

Orthographic Drawings. (See plan, section, and elevation.)

Overlay. A transparent sheet of paper, placed over an existing drawing, used to create an additional drawing with changes.

Paraline Drawing. Synonym for axonometric drawing.

Parallel Rule. A straight edge attached to a drafting board or table used to draft horizontal lines or as a resting place for other drafting instruments.

Perspective Drawing. A drawing of an element or concept giving the indication of height, width, and depth.

Photographic Print. A photograph made of a drafted plan.

Photostatic Print. A xerox-type reproduction of a drafted plan.

Pilaster. A portion of a rectilinear column set against a wall either for structural or decorative purposes.

Pine. A wood ranging from soft to moderately hard in nature and light tan to reddish brown in color, used in general construction.

Plan A drawing which is drafted parallel to the drafting surface and viewed at a 90° angle.

Plan Oblique. A drawing normally drafted at a 45° or 30°-60° angle. Such a drawing has a higher angle of view than an isometric, therefore giving the horizontal view more emphasis. All vertical lines remain vertical and all parallel lines remain parallel.

Plaster. A construction mixture composed of portland cement, sand, and water. Used to form wall covering.

Plywood. A sheet of wood composed of an odd number of layers of veneers joined with adhesive. Each veneer is placed perpendicular to adjoining veneers for added strength.

Pocket Door. Door that slides into a wall.

Poucher. Artistic method that can indicate material composition of construction.

Pounce Powder. A dry-cleaning powder, in bag form, used to camouflage smudges.

Prestype. Manufacturer's name for plastic film sheets capable of producing numbers, letters, or graphic patterns that can be applied to a drawing surface by pressure.

Protractor. An instrument used to draw curves of a specific degree.

Quadruplex Outlet. An electrical outlet with provision for use of four plugs.

Red Cedar. A soft reddish to brown wood usually employed in the formation of shingles.

Red Oak. A hardwood used for major construction, such as flooring. It is usually light to medium brown in color.

Redwood. A softwood ranging in color from light to deep reddish brown. It is used primarily for exterior purposes and furniture.

Reflected Ceiling Plan. A drafted drawing indicating the placement of all ceiling fixtures, beams, tiles, and HVAC's.

Rendering. The artistic embellisment of a perspective drawing or floor plan through the use of color and/or shading.

Riser. The vertical member of a step.

Sandblock. A lead sharpening device employing the use of sandpaper.

Sash. The horizontal member of a window that holds two panes of glass together.

Scale. The reduced measurement representing a larger measurement, e.g., ¼″ = 1′-0″, ⅛″ = 1′-0″, etc.

Sconce. A wall-mounted ligting fixture.

Section Drawing. An elevation-type drawing that cuts through an object exposing its interior construction.

Sepia Print. A reproduction of a plan on which corrections may be made.

Sill. Construction element usually forming the lower-most structure of an opening, e.g., door or window sill.

Single-Pole Switch. An electrical switch which operates from one location only.

Single Receptacle Outlet. An electrical outlet with provision for use of only one plug.

Sliding Door. Door that slides against a parallel wall or against another door.

Snake. A pliable rubber or plastic instrument used to draft curves.

Soffit. Underside of an overhanging cornice.

Softwood. Wood produced from coniferous, or cone-bearing, trees, e.g., cedar, fir, pine, redwood, and spruce. This term does not make any implication as to the strength or durability of the wood.

Spiral Stairs. Stairs that form a circle (often called a circular staircase).

Spruce. A softwood, pale yellow in color and used in general construction.

Straight Run Stairs. Stairs without interruption by a landing.

Structural Element. Any part of a design concept that may be considered architectural in nature, e.g., column, beam, window, or door.

Stud. Vertical support member of a wall.

Subflooring. The preliminary floor below the finished floor.

Technical Pen. A drafting instrument, with interchangeable points, employing the use of ink.

Template. A plastic sheet with predetermined shapes cut out in scale.

Three-Pole Switch. A switch which permits operation of the same fixture from two locations.

Title Box. That portion of a drawing that lists general information, such as client, date, scale, number, type of drawing, etc.

Tread. The horizontal member of a stair.

Triplex Outlet. An electrical outlet with provision for use of three plugs

T-Square. A drafting instrument, in the shape of a "T," used to draw horizontal lines or used as a guiding support for other drafting instruments.

Two Point Perspective. A drawing which depicts depth by having parallel lines converge to two vanishing points, thereby creating the illusion of depth. This differs from one point perspective in that it gives the viewer a wider and more interesting picture.

Two-Twenty Volt Outlet. An electrical outlet for use with heavier equipment, such as electric ranges, air conditioners, etc. Usually provides for grounding.

U Stairs. Stairs forming a "U" with a landing between each staircase.

Veneer. Thin sheets of wood used to form a false front on a lesser grade of wood. In addition, used in the construction of plywood.

Vellum. An excellent grade of drafting paper.

Wet Column. Any column encasing water pipes.

Wet Construction. Construction that consists of water and an additional mixture, e.g., plaster.

Winder Stairs. Stairs form an "L" shape without the assistance of a landing.

BIBLIOGRAPHY

Ching, Frank. *Architectural Graphics*. New York: Van Nostrand Reinhold, 1975.

Kicklighter, Clois E., and Baird, Ronald J. *Architecture: Residential Drawing and Design*. South Holland (Illinois): The Goodheart-Willcox Co., 1973.

Olin, Harold B., Schmidt, John L., and Lewis, Walter H. *Construction: Principles, Materials, and Methods*. Chicago: Institute of Financial Education, 1975.

Panero, Jules. *Anatomy for Interior Designers*. New York: Whitney Publishers.

Panero, Jules, and Zelnik, Martin. *Human Dimension and Interior Space*. New York: Whitney Library of Design (imprint of Watson-Guptill Publications), 1979.

ADDITIONAL READING WITH ANNOTATIONS

Anatomy for Interior Designers
Jules Panero
Whitney Publishers

An excellent book covering all the measurements of man in relationship to his environment for both commercial and residential projects. It is an invaluable tool when space planning is required. It covers all situations possible, such as sitting, standing, reaching, recreation, dining, sleeping, and movement requirements. It expands upon each room and the functional measurement requirements of each.

Architectural Graphics
Frank Ching
Van Nostrand Reinhold

An intermediate-level handbook with many illustrations. Topics covered include drafting tools, techniques, perspective drawings, elevations, and sections.

Architecture: Residential Drawing and Design
Clois E. Kicklighter
The Goodheart-Willcox Co., Inc.

An excellent and extensive reference textbook covering not only drafting techniques and presentation skills, but design and space allocation principles as well. In addition, it contains useful information covering construction materials, methods, costs, and career opportunities.

Construction: Principles, Materials and Methods
  Harold B. Olin, John L. Schmidt, and Walter H. Lewis
  United States League of Savings Association

Although this is not a drafting book per se, it is useful to the draftsperson because of its explanations and illustrations of construction materials and methods. As a rather large volume, it acquaints the draftsperson with the contracting field. In doing so, the draftsperson gains knowledge that would be most helpful in understanding contracting methods and would therefore be better equipped to graphically represent construction methods.

INDEX